Applied Cyberpsychology

Applied Cyberpsychology

Practical Applications of Cyberpsychological Theory and Research

Edited by

Alison Attrill
Senior Lecturer, University of Wolverhampton, UK

and

Chris Fullwood
Reader in Cyberpsychology, University of Wolverhampton, UK

APPLIED CYBERPSYCHOLOGY: PRACTICAL APPLICATIONS OF CYBERPSYCHOLOGICAL THEORY AND RESEARCH

First published 2016 by
PALGRAVE MACMILLAN

The authors have asserted their rights to be identified as the authors of this work in accordance with the Copyright, Designs and Patents Act 1988.

Palgrave Macmillan in the UK is an imprint of Macmillan Publishers Limited, registered in England, company number 785998, of Houndmills, Basingstoke, Hampshire RG21 6XS.

Palgrave Macmillan in the US is a division of Nature America, Inc., One New York Plaza, Suite 4500 New York, NY 10004-1562.

Palgrave Macmillan is the global academic imprint of the above companies and has companies and representatives throughout the world.

Hardback ISBN: 978–1–137–51702–9
E-PUB ISBN: 978–1–137–51704–3
E-PDF ISBN: 978–1–137–51703–6
DOI: 10.1057/9781137517036

Distribution in the UK, Europe and the rest of the world is by Palgrave Macmillan®, a division of Macmillan Publishers Limited, registered in England, company number 785998, of Houndmills, Basingstoke, Hampshire RG21 6XS.

Library of Congress Cataloging-in-Publication Data
Names: Attrill, Alison, editor. | Fullwood, Chris, 1976-
Title: Applied cyberpsychology : practical applications of cyberpsychological theory and research / Alison Attrill, Chris Fullwood, [editors].
Description: Houndmills, Basingstoke, Hampshire; New York, NY: Palgrave Macmillan, 2016. | Includes bibliographical references and index.
Identifiers: LCCN 2015039314 | ISBN 9781137517029 (hardback)
Subjects: LCSH: Internet—Psychological aspects. | Telematics—Psychological aspects. | Human-computer interaction—Psychological aspects. | Information technology—Psychological aspects.
Classification: LCC BF637.C45 A685 2016 | DDC 155.9—dc23
LC record available at http://lccn.loc.gov/2015039314

A catalogue record for this book is available from the Library of Congress.

A catalogue record for the book is available from the British Library.

Typeset by MPS Limited, Chennai, India.

Alison Attrill would like to thank David Attrill. Although no longer with us, his immeasurable support underpins my continued achievements. As always, my work is dedicated to my favourite big person, favourite little person, and favourite dog, for their continued love and support.

Chris Fullwood would like to thank Clive Ford; best-mate, band-mate, and sounding board for bouncing all of my best and worst ideas off of. I would like to dedicate this textbook to Mom (Sue Davies) and Dad (Michael Fullwood) for always believing in me and for helping me to realise my potential.

Contents

List of Figures and Tables

Figures

Table

Preface

Welcome to *Applied Cyberpsychology*. If you are reading this, you might already have an existing interest in the area of Cyberpsychology. If not, you might be wondering what this area of study is all about. This book revolves around the rapidly advancing area of psychology that considers all aspects and features of online behaviour. You might also come across the terms Internet Psychology, Human Computer Interaction, and Cybernetics. Whilst all of these play their own role in explaining and understanding the interface between humans and technology, this text centres specifically on the interactions of humans with what has become known as the world wide web (WWW). Although we use the word Internet to describe the WWW throughout this text, it is acknowledged that these are not the same thing. Whereas the Internet is the networking infrastructure that globally connects computers the world over, the WWW is a tool for accessing information over the Internet. You might think of the WWW as a network or web that sits on top of the Internet, and which allows us to create websites, search engines, links etc. in a way that disseminates information around or over the Internet. It is *the* Internet that transfers messages sent via email, but *an* internet (or intranet) that hosts a closed-group smaller network for a select number of users. Given the number of contributors to this text, you might find that these terms are used interchangeably throughout. You might now be asking yourself why we feel the need to create a text about the applied features of online behaviour, theory, and research.

One of the editors of this text recently gave a presentation at a yearly conference hosted by doctoral students of the University of Wolverhampton's Cyberpsychology Research Group (CRUW: www.wlv.ac.uk/cruw) on the need to consider online behaviour beyond the constraints of academic theory and research. Whilst us academics are working away to come up with theories that explain why, how, when, where, and what humans do on the Internet, we need to keep sight of the usefulness of those theories for everyday actions and behaviour. It is easy for academics to sometimes sit inside their ivory towers, without always fully considering how their research findings might impact the lives of real people. The psychology of offline behaviour has been considering real world applications for a long time, however online behaviour has only existed for the public for less than three decades.

The online world is constantly evolving and changing, sometimes for better, sometimes for worse. The mass media tends to highlight negative examples of Internet use, and it seems an artefact of psychological study to consider these behaviours in favour of positive interactions. In this text, we hope to draw attention to both sides of this coin, outlining some positive and some negative aspects of online behaviour.

In the Institute of Psychology at the University of Wolverhampton we also host a biennial international conference on all aspects of online behaviour. One thing that struck us when considering presentation submissions for the 2015 conference was the diversity of behaviour that we are trying to understand online. There is a tendency to see the WWW as a homogeneous platform, where people behave in the same way regardless of their individual characteristics, intentions, and the goals of their online behaviour. This notion has been well documented elsewhere (e.g. see Attrill, 2015a, 2015b), but needs revisiting if we are to understand human behaviour online. If you want to find a new life partner, offline you might go to a local pub or singles event. If, however, you feel ill and need medical advice, you would hopefully visit your doctor rather than a pub! Each of these behaviours is associated with diverse areas of psychological study: social psychology comes into play for the social roles and cues associated with these behaviours, whilst cognitive psychology might be considered in relation to the thoughts and emotions associated with the two different behaviours. Your online behaviour is no different. You might visit an online dating website to seek out a new partner, but consult the National Health Service website for information on your current medical symptoms. Why then, should we not consider all areas and aspects of online behaviour as being similarly diverse and guided by an array of psychological factors as offline behaviour? This text aims to further advance this diverse approach to understanding online behaviour by considering the applied features and aspects thereof. When perusing the contents table for this book, it might occur to you that we have omitted areas of theory and research that you would expect to see in such a text. When putting together any text to cover wide and varied areas of behaviour, there will undoubtedly be sub-disciplines that fall by the wayside. In this instance, we have left out at least one or two obvious applied areas of online behaviour, such as the impact of online interactions in educational settings. We thought it more useful at this juncture in Cyberpsychology to focus on emerging areas of interest and areas that require further theoretical consideration and research.

Given the rapidity with which the WWW has become an integral part of many people's lives, it is not surprising that we don't yet have a full appreciation of the psychology of many of its features. When we talk of theory and research being applied, we mean that it can be used to explain, guide, and/or understand actual behaviour, rather than providing a conceptualisation of behaviour. It is a consideration of the underlying or guiding psychological processes of people's interactions with others online. These can range from merely communicating with others via email and instant messaging, to engaging in criminal and deviant activities in cyberspace, or even shopping online. If you reflect on your own online behaviour, you will likely come up with a list of very diverse activities. Our aim with the applied focus of this book is to consider academic knowledge and understanding to make sense of these behaviours in a way that fosters further research, utilisation, and understanding by a wide reading public, including industry, service, and government providers. By providing industry and service providers with an understanding of the psychology underlying many online behaviours, diverse activities might be aided in the future. For example, the police might gain understanding that helps prevent online crime, whilst online retailers might consider the best layouts for their websites to promote inclusion, and online dating websites might deliberate on how best to protect their clients whilst offering a reliable and safe service.

Our considerations of online behaviour begin by looking at individuals and their online interactions. In the first two chapters, we examine the online behaviours of two specific groups in society and reflect on how we might foster safe, effective, and enjoyable use of the Internet for these individuals. In Chapter 1, the digital inclusion of individuals with disabilities is considered, whereas Chapter 2 offers a review of contemporary theory and research on parenting the online child. We then move on to consider cultural differences in online behaviour (Chapter 3), reflecting on intra- and inter-cultural variations in online conduct that might affect the offline and online lives of diverse peoples. All three of these areas could be considered at the macro or the micro level of study. In this instance, we explore some of the work that could be relevant to making individuals from varied socioeconomic, demographic, and dispositional backgrounds use the Internet to their advantage and disadvantage. We progress by assessing whether the Internet has a societal impact, by exploring digital inclusion and use across the lifespan (Chapter 4), and the role that the Internet might subsequently have on cognitive factors. More specifically, focus is given to the role of technology in memory (Chapter 5) as a cognitive example. This topic builds a nice bridge to subsequent chapters on the clinical applications of Internet usage. When we think of how the online word is portrayed in the mass media, panic around problematic use of the Internet is commonplace. Terms like *Internet Addiction* are habitually used on the assumption that the general public will understand the nature and consequences of excessive engagement with the online world. We therefore explore in detail a clinical perspective on Internet Addiction (Chapter 6), giving specific attention to issues surrounding classification and diagnosis. Our discussion then moves on to consider how the online world can be used as a supportive environment, both in the context of peer-support via online support communities (Chapter 7) and in the ways in which counselling psychologists can make use of the online world to engage with their clients (Chapter 8). In both of these chapters we reflect on the positive and negative implications of supporting and counselling others online, and offer recommendations for best practice. One area of online behaviour that receives an enormous amount of press coverage, as well as research attention, is that of interpersonal relationships online. The ways in which research can inform the role that online dating might play in these interpersonal relationships (Chapter 9) is therefore subsequently considered. Given that online dating has become a multi-billion-pound income generator, further uses of Internet psychology are explored in relation to online consumer behaviour in Chapter 10. The final four chapters of this text use more specific examples of applied theoretical approaches. In Chapter 11, the role of psychology in developing games is considered, followed by the specific application of certain theoretical models and research to aiding military understanding and practice in Chapter 12. Chapters 13 and 14 revolve around organisational psychology. We first consider the ways in which social media are impacting on, and are impacted by the use of the Internet, prior to exploring the specific use of a multitude of tests for assessing individuals, both for their suitability for given tasks and for their execution of roles and goals, in Chapter 14.

We hope that you enjoy this text and that it sparks in you an interest in exploring further some of the areas of online behaviour covered.

References

Attrill. A. (Ed.) (2015a). *Cyberpsychology*, Oxford, UK: Oxford University Press.
Attrill, A. (2015). Palgrave Studies in Cyberpsychology. In *The Manipulation of Online Self-Presentation: Create, Edit, Re-edit and Present*. Basingstoke, UK: Palgrave Macmillan.

Notes on Contributors

As one of the UK's largest Cyberpsychology research groups, we were very fortunate in the creation of this book to have so many contributors readily available in our midst to contribute chapters. Where chapters have been authored by academics outside of our institution, the majority have been affiliate members of our group for some time, often attending our events and collaborating on world-wide research projects. We would like to thank the following contributors for taking the time to offer their expertise to write for this text. For further information on any of the contributors, please refer to their institutional website in the first instance.

Edward T. Asbury is Associate Professor of Psychology at Texas Woman's University (TWU) in Denton, Texas. He teaches cyberpsychology, developmental psychology, applied statistics, positive psychology, and physiological psychology. Prior to his tenure at TWU, he worked as a family therapist and a research analyst for a public school district. His research interests centre around social networking sites for at-risk populations, and technology and boundary management.

Alison Attrill is Senior Lecturer in Psychology at the University of Wolverhampton. She is a joint coordinator of the Cyberpsychology Research Group and specialises in applying psychological theory and explanation to a wide variety of online behaviours, most notably to understanding how people create and share different versions of their selves online.

Nick Banks is Senior Lecturer in Counselling Psychology at the University of Wolverhampton. He is a Chartered Clinical Psychologist and has a particular interest in counselling and culture.

Darren Chadwick is Reader in Applied Psychology associated with the Wolverhampton Intellectual Disability Network (WIDeN) and the Institute of Psychology, at the University of Wolverhampton, England. Darren has made use of quantitative, qualitative, narrative and participatory and inclusive methods to explore the well-being, communication and life course of people with a learning disability. Some of his recent research work has been within the field of cyberpsychology, focusing on the digital divide, online identity and the benefits and risks experienced by people with intellectual disabilities in the online world.

Coral Dando is Professor of Cognitive Psychology at the University of Wolverhampton, Chartered Psychologist and Scientist, and a Consultant Forensic Psychologist (HCPC). Her primary research interests are centred on cognition in interviewing contexts, with reference to contemporary theories of long-term memory, social cognition, and cognitive load. She has worked extensively with the UK and US governments, and other organisations such as the International Criminal Court and United Nations High Commission for Refugees. Her most recent research projects include modelling and investigating insider threat, supporting accurate eyewitness remembering, and developing and evaluating psychologically informed, goal-directed, conversational-style interview techniques for intelligence interviewing to maximise the detection of verbal deception across various security settings.

Nicola Derrer-Rendall is a senior lecturer at the University of Wolverhampton, UK. She completed her MSc in Work Psychology and Business at Aston Business School, Aston University, UK, and is a chartered member and Associate Fellow of the British Psychological Society. Her current research interests include confidence and its impact on performance, individual goal-setting and achievement, and personality and academic confidence.

Nicola Fox Hamilton holds an MSc in Cyberpsychology from the Institute of Art, Design and Technology, Dun Laoghaire (IADT), and is completing a PhD at the University of Wolverhampton. Her current research examines the connection between language, personality, and attraction in online dating profiles. Nicola is Lecturer in Cyberpsychology in IADT, and is co-chair of the Psychological Society of Ireland's Special Interest Group for Media, Art, and Cyberpsychology (SIGMAC).

Chris Fullwood is Reader in Cyberpsychology at the University of Wolverhampton, UK and joint coordinator of the Institute's Cyberpsychology research group. He completed his degree and PhD at the University of Stirling, Scotland. His current research interests include personality types in cyberspace, blogging motivations and behaviour, online impression management, and the impact of technology on the social and educational development of young people.

Rachel Harrad is a PhD student at the University of Wolverhampton, with a background in both adult nursing and psychology. Her research focuses on cyberpsychology, together with aspects of self-evaluation and engagement with social media.

Daniel Hinton is a lecturer in the Institute of Psychology at the University of Wolverhampton, specialising in Occupational Psychology. His research is focused on psychometric testing, particularly the ongoing issue of ethnic differences in ability test performance. Through his consultancy, he has worked with a diverse range of clients both within the UK and internationally, developing psychometric tools for clients in both the public and private sectors.

Linda Kaye is a senior lecturer in the Department of Psychology at Edge Hill University. Her primary research interests surround the social contexts of digital gaming and their impact on psychosocial outcomes, such as self-esteem and psychological well-being. She also has interests in the role of identity in virtual contexts and the extent to which identity processes account for variations in psychosocial outcomes.

Daria Kuss is a chartered psychologist, Senior Lecturer in Psychology, and a senior researcher at the International Gaming Research Unit at Nottingham Trent University, UK. Her research area encompasses media addiction in general and Internet, gaming, and mobile phone addiction specifically. She is interested in the Internet in gaming use, associated psychopathology, behaviours, and personality dimensions. Her research activities encompass various theoretical and empirical research projects in collaboration with colleagues across the world.

Tom Mercer is a lecturer and cognitive psychologist based within the Institute of Psychology at the University of Wolverhampton. His main research interest is memory, and current projects are examining the factors responsible for forgetting. His other research is investigating different techniques that can be used to reduce memory loss.

Johanna Myddleton is a recent psychology graduate of De Montfort University who is studying for her PhD at the University of Wolverhampton. Her primary area of research is cyberbullying in the workplace, and she has a particular interest in people's perceptions of online behaviour and the perpetration of online anti-social behaviour.

Sally Quinn is a teaching fellow at the Department of Psychology, University of York. Her main research interests lie in the social uses of technology with a specific focus on the positive uses of the Internet and mobile technology. Her work has appeared in *British Journal of Developmental Psychology, Computers in Human Behavior, Mobile Media and Communication,* and *Cyberpsychology, Social Networking and Behavior.*

Debbie Stevens-Gill is a senior lecturer in the Institute of Psychology at the University of Wolverhampton and course director of the accredited MSc in Occupational Psychology. Her primary research interests involve stress and well-being at work. In particular her research interests include coping, emotional regulation strategies, and resilience. She is a registered occupational psychologist who specialises in psychological assessment, including training test users to British Psychological Society (BPS) qualification standards in occupational testing, though areas of expertise also include organisational development, work design (including motivation and human factors), and careers coaching.

Claire Tranter is a PhD student at the University of Wolverhampton. Her research focuses on the effects of culture, cognitive style, and context on persuasion outcomes. She is examining theoretical and applied understandings of social cognition, and how judgments are formed and modified to maximise information gain for intelligence interviewing purposes across traditional face-to-face, synchronous and asynchronous textual, and virtual environments. Her research programme is funded by the UK Ministry of Defence; her Director of Studies is Prof Coral J. Dando (with Dr Chris Fullwood and Dr Darren Chadwick as second supervisors).

Caroline Wesson is Senior Lecturer in Psychology at the University of Wolverhampton, where she teaches on social psychology and forensic psychology modules. She also supervises students on the counselling psychology doctorate programme. Her research interests focus on intellectual disability and online behaviour, the interface between people with intellectual disabilities, mental health, and the criminal justice system, and the application of confidence to applied settings (e.g. within educational, training, and forensic contexts).

1
Digital Inclusion and Disability

Darren Chadwick and Caroline Wesson

1.1 Contextualising digital disability

In this chapter, we will apply cyberpsychological, psychological, and sociological theories to the issue of inclusion in the online world for people with disabilities, a group of individuals often overlooked within society. The use of technology to communicate has become an essential and socially acceptable aspect of most people's lives and it is becoming increasingly difficult to distinguish between the "digital world" and the "real world" (Helsper, 2008; Ritchie & Blanck, 2003). Hence, Digital Inclusion is an increasingly important social issue, reflecting imperatives, opportunities, and considerations about human rights, equity, issues of identity, language, social participation, community and civic engagement, and opportunity pertaining to the digital world (Castells, 1997; Warschauer, 2003).

Before we continue, imagine for a moment how you might feel if you noticed that everyone else was using new technologies and methods of finding things out and communicating with each other, but you were excluded from these activities. What if you found that it was difficult to understand how to use a computer, tablet or a smart phone, that the buttons were too fiddly and the interface when it was turned on was incomprehensible, that everything was written in a language you had never learned, or that other people told you that you weren't allowed to use these technologies because they were too risky or you might break or lose them? Consider how those experiences might make you feel. Research has indicated that these scenarios reflect the real lived experiences of people who are disabled by society. Herein we will outline research that has provided insights into the factors that influence the inclusion of people with disabilities.

1.1.1 The possibility and promise of the digital world

People with disabilities are considered likely to be able to make especially fruitful use of the online world to help overcome disabling barriers they face caused by societal attitudes, organisation, and structuring which mean that their differences, for instance physical, sensory, intellectual or psychological impairments, are not adequately considered and so they are discriminated against. For example: a person with considerable physical impairments can pursue online education meaning there is less need to leave home; a person with a significant visual impairment can gain access to documents by downloading them and converting the text to speech; a person with a learning disability can socialise and make or maintain friendships from home. Sadly though, disabled people, because of poverty, lack of social support or other reasons, frequently lack the means to get online and if they can, may not be adequately equipped or supported (Chadwick et al., 2013b; Hoppestad, 2013).

1.1.2 The digital divide

The exclusion of people with disability from the online world has been referred to as one important component of the "digital divide" (Dobransky & Hargittai, 2006; Warschauer, 2003). Those with disabilities have been classified as socially isolated in digital terms (Helsper, 2008). The deep social exclusion experienced by many with disabilities is exacerbated by a combination of limited educational opportunities around Information Communications Technology (ICT), low income, unemployment, comorbid health problems, and low social status and has been implicated in a concomitant digital exclusion involving limited access to and use of the Internet (Helsper, 2008). For example, in 2013 in the UK 31% of disabled adults had never been online, representing over half of all the people in the country who have never used the Internet. Thus, those with disabilities, within this context, appear to be the largest group of people excluded from the online world (ONS, 2013). Research from other countries has supported the lack of access experienced by people with disabilities (e.g. Palmer et al., 2012). However, a recent increase amongst younger people with disabilities has also been reported (e.g. see Feng et al., 2008). Nevertheless, it is doubtless that, comparatively, digital exclusion persists for disabled individuals.

1.1.3 Defining disability

People with disabilities are an extremely heterogeneous group, with complex aetiologies, presentations and often comorbidity. For example, for those with intellectual disabilities (ID), 20–30%, it is estimated, will also have physical disabilities, and 10–33% will also have sensory impairments (Hatton & Emerson, 1995; McLaren & Bryson, 1987). Clinical definitions of disability remain deficit-focused, but these have been moderated in recent years with an acknowledgement that disability is a socially constructed term, culturally and historically grounded, that is used to label a particular group of people within society (Goggin & Newell, 2003; Manion & Bersani, 1987).

Alongside this there has been a move towards acceptance, tolerance, and inclusion, with significant efforts since the 1980s, occurring largely in the global north, to remove the societal barriers disabled people experience (Brown, 2007). More recent thinking around disability indicates that the identification of deficits should be integral to the identification of necessary support people require to overcome these challenges and that people's strengths should also be highlighted (Schalock et al., 2010).

This view fits with a more social or interactionist model perspective on disability. Briefly, the social model of disability views disability as the restrictions imposed on a person due to societal attitudes, organisation and structures which lead to the world being organised in a way that excludes and discriminates against people who differ due to natural variations in terms of their physical, cognitive, sensory, and psychological functioning, also termed impairments (Oliver, 2013). This approach contrasts with the hegemonic medical model, whereby disability is instead viewed as the limitations a person faces due to the impairments they have. In the social model the responsibility for restrictions and exclusion is said to be with society, but with the medical model the issue dwells in the individual's impairments, leading to the desire to remove rather than accept or accommodate impairment and difference (Shakespeare, 2014).

The interactionist perspective is linked to the social model but takes a more pragmatic stance, acknowledging environmental factors and how these interact with specific and unique personal factors (e.g. impairments). It also recognises the role the impairment can play in disabling the person (e.g. pain or profound cognitive impairment) (Shakeseare, 2014). It is the combination of these intrinsic (personal) and extrinsic (social and environmental) factors which creates barriers and exclusion for the person, or conversely creates strengths that facilitate inclusion (personal communication, Buell, 2015). The interactionist perspective has been conflated with the bio-psycho-social model but differs in orientation, in that the bio-psycho-social model takes a deficit focus of disability, where disability is primarily a function of the impairments rather than due to societal factors. The interactionist model, like the social model, situates the problems disabled people face within society. It is the interactionist model that provides the theoretical underpinning to this chapter. In the next section we consider the opportunities and ways in which the digital world can and has been found to enhance the inclusion of people with disabilities and the associated benefits this can bestow.

1.2 Enhancing inclusion via the digital world

As noted earlier, technological advances and the continuing growth of the digital world may enhance the inclusion of people with disabilities, having the potential to help them better integrate into society and experience many of the benefits of full citizenship that most take for granted (Foley & Ferri, 2012). Using the Internet could afford opportunities for people with disabilities to access information and engage in social interactions that could be difficult in the offline world by decreasing or removing the barriers that may prevent them from participating in such activities (D'Aubin, 2007; Dobransky & Hargittai, 2006; Hoppestad, 2013).

Vanden Abeele et al. (2012) argue that the dependency some disabled people may have on others (e.g. for mobility, information, social interaction) can potentially be transferred to the Internet. Indeed, Internet access may offer learning and education opportunities, access to employment, entertainment, self-expression, and social, political, economic, and cultural networking, connectedness, and inclusion (*cf.* Chadwick et al., 2013b; Stendal, 2012). Engagement with the online world may be within the context of people's disability or completely separate from this, with individuals having the freedom to disclose, or not, their disabled identity at will (Bowker & Tuffin, 2003; Cromby & Standen, 1999; Thoreau, 2006). Consideration will now be given to evidence pertaining to some of these opportunities as potential means of enhancing inclusion.

1.2.1 Online social inclusion, relationships and social capital

The online world is having an impact on social relationships, providing new methods of connecting and engaging with other people that have yet to be fully explored and understood (Vallor, 2011). One way of conceptualising the social inclusion opportunities afforded by the digital world is via Social Capital Theory, which posits that, through the processes of strengthening and maintaining social ties, one can enrich existing bonds and develop new social ties. Theorists here have highlighted that social relationships are primarily reciprocal in nature, being a route to mutual social benefits and mature relationships (i.e. Bourdieu, 1983; Coleman, 1988; Putnam & Goss, 2002). Researchers have considered the model when investigating the valence of the Internet and social media in facilitating social capital ties (Valenzuela et al., 2009). Social capital enrichment and bonding may be enhanced and sustained by social media (Williams, 2006).

Research taking a Social Capital perspective with typically developing people, has indeed indicated that young people are motivated to join social networking sites to maintain strong friendship ties and to strengthen new ties, but are less motivated by the opportunity to meet new people online (Ellison et al., 2007). Ellison et al. (2007) went on to argue that "maintained" social capital expresses how social networking sites, Facebook in this instance, helped to maintain contact between pre-established older friend networks. Thus people share information online to help sustain, enrich, and strengthen offline friendships and relationships. This idea has been supported in a large scale study of young people without disabilities across Europe (Livingstone & Haddon, 2009).

There is increased probability that social relationships and opportunities for social interaction will be impoverished amongst people with disabilities (e.g. Batorowicz et al., 2014; Forrester-Jones et al., 2006). Despite this, people with disabilities share similar social aspirations to their typically developing peers (e.g. Garcia et al., 2014; Soderstrom, 2009), wanting both romantic and platonic relationships. With regard to online romantic relationships, researchers have begun to explore aspects of online dating (Finkel et al., 2012), but despite the existence of long-lived dating sites for disabled people (e.g. Dating4 Disabled.com and whispers4u.com) empirical research on this topic appears scarce.

Moving on to friendships, physical and/or communicative difficulties interacting with social and environmental barriers may restrict some individuals with disabilities from developing friendships, in turn contributing to feelings of loneliness (McVilly et al., 2006). Access to the Internet can go some way towards overcoming this, as it may significantly improve the quality and quantity of disabled people's social interactions (Guo et al., 2005). Individuals who are unable to work due to their disability have been found to be more likely to use the Internet for social interaction than those who are employed (van Deursen & van Dijk, 2014), with the Internet presumably providing them with some level of social interaction that employed individuals might enjoy in the workplace.

The Internet also brings with it the opportunity to be part of online communities consisting of similar others. Gillespie-Lynch et al. (2014) found that individuals with autism spectrum disorder felt that the main benefits of computer-mediated communication were the opportunities to practice social interaction and engage with similar others online. The sense of belonging that this may afford can lead to a reduction in feelings of isolation and an increase in the availability of social support (Obst & Stafurik, 2010). For example, online support groups may provide a comfortable and supportive environment in which to seek advice and support and discuss personal issues with people with similar conditions (Fullwood & Wootton, 2009). Third and Richardson (2009) also discovered that young people with a chronic health condition or disability found that social networking in an online community was positively supportive of their health and well-being. Prior research has confirmed the relationship between social support and well-being and has suggested that support networks can provide protection against stress (e.g. see Cohen & Wills, 1985), and access to the Internet has the potential to ultimately enhance well-being for people with disabilities (Goggin & Newell, 2003; Obst & Stafurik, 2010). Kirk (2008) reported that young people with complex healthcare needs and physical disabilities often find it difficult to keep in touch with friends and acquaintances as they travel through life with points of transition (e.g. moving home, school etc.) sometimes meaning the end of valued relationships. However, the use of mobile communication technologies has been found to increase societal participation by supporting social networks for Augmentative and Alternative Communication (AAC) users (McNaughton & Bryen, 2007).

Although the Internet allows for people with disabilities to connect with similar others it is also an opportunity to expand one's interactions. Rather than solely being about "maintained" social capital, Holmes and O'Laughlin (2012) found that people with ID preferred to access social networking sites that allowed for interactions with a wider audience rather than one restricted to others with ID (as people with ID are perceived to be vulnerable and therefore safer social networking environments have been created). Similarly, online communication allows deaf/hearing impaired individuals to engage in group discussions with hearing individuals, an activity their impairment makes difficult in the offline world (Vanden Abeele et al., 2012). The Internet also provides an opportunity for a private life, away from carers (Löfgren-Mårtenson, 2008), giving the individual the freedom to decide where they visit (in terms of websites) and with whom they interact.

Underpinned by Social Capital Theory, Hynan et al. (2015a) developed a grounded theory of Internet and social media use from interviews with 25 people with cerebral palsy who used AAC devices. The conceptual theory showed that young people using AAC desired to use the Internet and social media, with some reporting that they would feel lost without it, and participants were thankful for the collaboration they had (i.e. working jointly with a communication partner to create, modify, access, and share content online when they were not able to do this independently due to cognitive and/or literacy limitations and/or physical access barriers).

For these people, desire and self-motivation to use the Internet was fostered via technological innovations and access methods, prior experience and human support from family, friends, and educationalists. The digital skills of, and restrictions imposed by, these supporters were important influences on access and online inclusion. Also important were requests to access the Internet and preferences made by the individual, though such requests were not always successful. Social media sites (Facebook, Skype, and YouTube) were those predominantly used by participants, with sites viewed as less social (e.g. Twitter) less often used. Access was primarily via a desktop computer as mobile technology use was extremely limited. Switching between AAC use and computer use reportedly slowed communication for participants.

The grounded theory illustrates how desire to use the Internet and social media was predicated on perceived opportunities to enrich social relationships and to enhance self-presentation and self-determination (Hynan et al., 2015a, 2015b). A recent systematic review of 10 studies by Caton and Chapman (In Press), investigated the use of social media by people with ID. Nine themes were identified, demonstrating parallels with Hynan et al.'s studies, regarding the importance of social identity, relationships, support and ultimately, happiness and enjoyment in the use of online social media. In a subsequent section we will further consider the interface between self-expression of identity and stigma and the digital world, but next we will present a theoretical explanation of motivation to use the Internet.

1.2.2 Motivation to be included in the online world

The research discussed previously indicates that people across a broad spectrum of impairments and disabilities are motivated to use and engage in the online world, enjoying the benefits offered, in particular how it can facilitate and enrich relationships (e.g. Hynan et al., 2015a; Caton & Chapman, In Press). The uses and gratifications approach (Katz et al., 1974) provides an explanatory framework for why individuals may be motivated to engage in the use of mass media to meet their psychological needs. Applied to computer-mediated communication Ruggerio (2000, p. 26) classifies these needs into four typologies:

1. Diversion (e.g. escapism/emotional release)
2. Social utility (e.g. information acquisition for conversations)
3. Personal identity (e.g. reinforcement of attitudes, values, and beliefs)
4. Surveillance (e.g. learning about one's community, events, and activities)

Examining blind/visually impaired and deaf/hearing impaired individuals' Internet use from a uses and gratifications approach, Vanden Abeele et al. (2012) found that the most common usage of the Internet amongst these groups was information-seeking and communication with friends/family. Whilst this may gratify the needs Ruggerio outlines, these gratifications were all underpinned by the need for independence and active agency. By accessing the Internet, individuals were able to find out information for themselves (e.g. bus timetables) and then make arrangements (e.g. to visit friends) as a result of this information. For people who are often reliant on the help of others, this ability to organise one's own life makes the Internet an empowering agent (Vanden Abeele et al., 2012).

However, Vanden Abeele et al. (2012) warn that access to the Internet is not unproblematic in terms of its inclusive benefits, with a discrepancy apparent between gratifications sought and obtained. Specifically, because disabled people may have a higher reliance on the Internet than non-disabled, this can lead to a constant feeling of "catching up" (in terms of keeping up to date with technological advances) in order to keep on benefitting. Thus whilst the Internet can be liberating it can also become a power-relationship, with the disabled person forming a dependency on the Internet; but this potential risk has not be empirically explored.

So, whilst the Internet has great promise with regards to the inclusion of people with disabilities it is not a panacea. The inclusive benefits afforded online may not extend to the offline world and strategies to enhance inclusion should consider intrinsic *and* extrinsic factors and their interaction, so as not to create a reliance on one approach. However the online world potentially has unique and unparalleled benefits for people with disabilities, as explored in the next section.

1.2.3 Inclusion, identity and stigma

A key benefit to inclusion to going online is afforded to those with a visually identifiable impairment by the very nature of the computer-mediated environment: its visual anonymity. Such anonymity could potentially reduce the stigma an individual may otherwise experience. Whilst not all online communication is visually anonymous (e.g. one may post photos/videos online), it remains predominantly textually based. Thus disabled people may mask differences if they so choose, which has the potential to reduce any stigma they may experience (Cromby & Standen, 1999). In terms of social inclusion and relationship development this may impact greatly, as physical appearance can affect first impressions and relationship formation (Park, 1986). Anonymity can thus impact how others view the individual (impression formation) and how individuals view themselves (flexible identity construction) (Chadwick et al., 2013a), and may ultimately support inclusion and self-defined identity presentation by preventing exclusion based on initial visual impressions. However, the claim that the Internet is an emancipatory space for disabled people is not an uncontested idea (Thoreau, 2006), and it has been posited that hiding disability is likely to do little to reduce stigma or increase acceptance (Bowker & Tuffin, 2003).

Considering the social construction of disability, as forwarded in the interactionist model, this highlights the relationship between technology and disability. Vanden Abeele et al. (2012) argue that technologies are not just instruments to overcome physical impairments which may serve to facilitate inclusion; they may also simultaneously serve to exclude and reinforce disabled identity. For example, in the case of a speech output AAC device, the visual cue or nature of speech output may serve to highlight that the individual using it is disabled.

1.2.4 Educational inclusion

In addition to social inclusion, developmental, civic, and political inclusion may also be influenced by the digital revolution. Although it is outside of the scope of this chapter to cover all of these, we have touched upon these aspects of inclusion throughout the chapter, and here focus specifically on one aspect which, though primarily developmental in nature, has the potential to have subsequent impacts on both civic and political inclusion, namely, e-learning.

E-learning, on the surface, provides the opportunity for all students to be equal, with disability being rendered invisible. This potentially means that the disabled student may interact with their online cohort to the same extent as the non-disabled student, whereas in traditional educational settings physical and communicative barriers may exist to prevent this. In the UK alone, 10% of students in 2103/14 were identified as having a self-reported disability (Higher Education Statistics Agency, 2015), although the actual figure may be higher as students are not obliged to report a disability. Online virtual environments can be used for education and skill acquisition by people with disabilities (Kent, 2015), with the Internet found to be successful in reducing physical barriers to education and learning (Guo et al., 2005).

Despite the potential benefits of e-learning, 46% of students that Roberts et al. (2011) surveyed indicated that they felt their disability was a barrier to success on their online course. This may partly be because failure or reluctance to disclose a disability may lead to a lack of suitable adaptations or special arrangements being made to accommodate disabilities. Indeed Kent (2015) argues that there is the risk that students with disabilities may become invisible online. Whilst e-learning platforms have great potential for inclusivity, they are not always accessible (Fichten et al., 2009; Kent, 2015). Foley and Ferri (2012) argue for technology for people, not disabilities, in terms of technology being made accessible from the outset rather than adapted later. However, accessibility is often not a factor that is considered when digital resources and online courses are designed (Bunning et al., 2010; Dawe, 2006; Hollier, 2012; Palmer et al., 2012; Roberts et al., 2011; Wehmeyer et al., 2004).

To summarise then, there is evidence supporting some of the many theorised benefits that the online world can bestow. Though some consistency around social motivation and benefit is emerging (e.g. social and identity aspects) for other areas, research findings are lacking or are less consistent (Cheatham, 2012), with studies often reporting both benefits *and* barriers (Dobransky & Hargiatti, 2006; Drainoni et al., 2004). Thus there is a need for more research in this area and it is evident that digital exclusion for many people with disabilities remains a reality. Evidence does not extend to all benefits or disability groups due to barriers with regards to both access to and effective use of the Internet by those with disabilities.

A full discussion of barriers to digital exclusion is outside the scope of this chapter. However, utilising the initial interactionist framework outlined earlier, we can characterise barriers to digital inclusion as either intrinsic (e.g. cognitive, physical, psychology, and sensory impairments) or extrinsic (e.g. societal and carer attitudes, lack of support and training, financial and economic factors, and lack of policy and governmental support for ICT accessibility) to the person, with the interaction of these factors fostering digital inclusion or exclusion (*cf.* Chadwick et al., 2013a; Chadwick et al., 2013b; Dobransky & Hargittai 2006; Hoppestad, 2013).

1.3 Online risks for people with disabilities

Despite the extensive discussion, digital benefits are not the whole picture and inherent risks for disabled people online have yet to be adequately studied and represent another potential contributor to digital exclusion. Online risks include engagement in antisocial behaviour (e.g. illegal downloading, bullying, uploading sexually inappropriate pictures or text), negative contact online (e.g. having personal information stolen, being bullied, being groomed, unwelcome persuasion) and exposure to harmful, manipulative or exploitative content (e.g. advertising, violent or hateful material, harmful sexual material, extremist or racist information) (Livingstone & Haddon, 2009). Recent research has indicated that the general public perceives both the risks and benefits of being online to be greater for people with ID when compared with themselves (Chadwick et al., 2015b). Other risks that may also be perceived as greater include the likelihood that people will damage technology or be the victims of crime (e.g. breaking or having their smart phone stolen).

A core narrative around people with disabilities is that of the vulnerable, eternal child (Pueschel & Scola, 1988). Those supporting people with disabilities to go online may act as gatekeepers, preventing access and inclusion raising the issue of protection, control, and what it means to be a self-determining member of society. Physically disabled adolescents have been found to be more likely than their non-disabled peers to be warned, by their parents, of the risks of cyberspace and to have restrictions placed on their Internet use (Lathouwers et al., 2009; Raghavendra et al., 2012). Volunteer training of people with ID to stay safe online found that those providing support took control over decisions regarding what was safe and acceptable online content for people with ID to access (Seale, 2003). Thus it appears that disabled people are more likely to be viewed as in need of protection when online. Such societal and carer views of people with disabilities as more vulnerable and less able to access the Internet are likely to lead to increased digital exclusion via reduced access, support, and training (Chadwick et al., 2015b; Livingstone & Haddon, 2009; Seale, 2014). As Seale et al. (2012) posit, perceived vulnerability, issues of safeguarding, and concerns about performance can govern the ways in which supporters of people with disabilities consider and implement risk management. Clearly, acknowledging and addressing the attitudes and concerns and enhancing the resources and skills of carers regarding ICT use requires further exploration and investigation (Palmer et al., 2012).

1.3.1   Positive risk-taking and the online world

Concerns that people with disabilities may be at great risk online do not currently have much empirical support (see Plichta, 2011; Seale, 2014). For instance, Didden et al. (2009) found a reduced likelihood of children with ID being cyberbullied (7%). Despite being perceived to be at greater risk (e.g. Chadwick et al., 2015b) this is not well supported for disabled people in the existing empirical evidence. A number of possibilities exist that may explain this and require further exploration. It may be that risks are more likely to be experienced by people with disabilities but that the barriers outlined above that reduce access also serve to reduce the likelihood that online risks will become a reality. It could be that being able to hide disability online may reduce the likelihood of some risk, e.g. being bullied, groomed for sexual exploitation, or scammed. People with disabilities may access less risky online spaces (e.g. advocacy and support groups that are disability-specific) or may be monitored more to prevent accessing potentially more risky sites. Those who access the online world with disabilities may be more capable, resilient, and well-supported and so online risks are obfuscated or avoided more. These hypotheses remain in need of empirical investigation.

A question raised by Seale (2014) in consideration of this, adapted by the authors, is how can those with disabilities and those who support them negotiate and balance the benefits and opportunities of being online against the potential risks? Seale (2014) offers the lens of positive risk-taking as a possible route to enhance understanding. Positive risk-taking involves enabling people with disabilities to have greater control over the ways in which they choose to live their lives.

Promoting well-being through independence and choice, the decisions made may involve risk-taking, for example in terms of health, safety, or potential failure and disappointment but is pursued nonetheless. Risk is not avoided, but potential harm is weighed up as being less important than the potential benefits of success, and risks are acknowledged and managed (Alaszewski & Alaszewski, 2002; Morgan, 2004). Within this framework risk is seen as a shared activity, necessary for growth, learning, and development, and an integral part of the human experience (Perske, 1972), hence it necessarily involves shared decision-making and negotiation. Recent qualitative research has found that adults with ID who experience online relationship and financial risks recounted that they had learned from the experience and would better manage similar future situations, suggesting the potential for post-traumatic growth and resilience-building through experiencing online risks (Chadwick et al., 2015a).

Löfgren-Mårtenson's (2008) qualitative study revealed a discrepancy between the views of the young people with ID interviewed and those of support staff regarding the advantages and disadvantages of being online. The young people had a more positive view of the Internet in contrast with the primarily pessimistic view of the staff who, by and large, saw the Internet as being unsuitable for people with ID. This does not necessarily bode well for the use of this positive risk framework in future support without the development of further guidance. As Seale (2014, p. 222) notes, there is "a need for more research into how issues of risk, safety and support influence the quality of technology use by people with learning disabilities." We would argue that this need is also likely to be useful for others with different impairments. Strategies to aid negotiation of online risk are just one approach to enhancing digital inclusion and in the next section we consider some of the other strategies that have been employed.

1.4 Online inclusion strategies

1.4.1 Social networking programmes

A number of strategies have been employed to encourage Internet use amongst groups of disabled individuals. For disabled people, existing social capital can facilitate online inclusion. Home Interventions have been devised, involving support, training, and introduction of assistive technology to enhance online inclusion of adolescents with cerebral palsy, physical disability or acquired brain injury, to enhance development of social networks and in turn reduce loneliness. Whilst training did enhance social networks by increasing the number of online partners the young people had, it did not lead to an overall significant reduction in loneliness. However, it is important to note that baseline measures indicated that this group were not very lonely to start with (Raghavendra et al., 2013). Raghavendra et al. (2012) also found that friends and siblings played a significant role in supporting young people with physical disabilities to get set up on the Internet and to use social networking and other sites.

Magee and Betke (2013) also developed a system aimed at young people with physical disabilities to automatically generate messages to provide answers to the question "What did I do today?" which goes on to post these as updates on social networking sites as support for social networking activities and to enhance online communication with family, friends, and carers. Findings indicated that such interventions can prove successful in enhancing online social engagement.

1.4.2   E-mentoring

E-mentoring has been posited as providing an opportunity to provide support, exchange practical information, and experience an accepting relationship. Shpigelman et al. (2009) investigated one such programme designed for use with young people with special educational needs. Social and emotional support was provided by disabled mentors for disabled protégés who were 15–20 years of age. Qualitative findings suggested that the programme had positive outcomes for the personal development and empowerment of the young people. Shpigelman and Gill (2013) went on to outline unsuccessful aspects of the e-mentoring programme to help inform future work of this kind, suggesting that more informal and supportive communication was viewed as more successful compared with more formal communication with a more distant tone. It would appear that as a support mechanism this has potential, though wider application and evaluation is required.

1.4.3   Improving accessibility

As noted earlier, the interfaces of ICT and associated programmes and applications can be off-putting, challenging and extremely difficult to use for those with sensory, cognitive, and physical disabilities. Some headway is being made in producing an "accessibility layer" that sits on top of existing programmes and application interfaces, which aims to make them easier to understand for people with disabilities (e.g. "able to include") by converting text to easily recognisable pictograms across different European languages (Sevens et al., 2015). The ultimate goal of the project is to enhance independence and citizenship (Tilly, 2015). Preliminary findings are promising but the long term and wider utility of this approach is as yet untested. Sundqvist and Ronnberg (2010) conducted an intervention study aimed at supporting children to use symbol-based email, which was found to help alleviate unequal turn-taking and reduce time pressure, so there does appear to be some promise here too.

1.4.5 Virtual environments

In contrast with the discussion of risk earlier, virtual reality is frequently perceived as offering a safe environment in which to teach skills that are associated with some level of danger (e.g. pedestrian safety and stranger safety) or a forum in which to engage in leisure, social networking, and try out real world skill development (Cromby et al., 1996; Stendal et al., 2012). Despite the assumption of benefit that practice in virtual worlds may offer, implicit in the way virtual reality is used is an assumption that trying things out in the real world is potentially too risky and hence is to be avoided (Parsons et al., 2000). Stendal et al. (2013) found that people with lifelong disability perceive that they reach larger and more diverse networks via virtual worlds, suggesting that these types of digital environments have been viewed as beneficial. Although the potential for virtual environments, as yet, appears not to have been fully tapped, consideration is needed as to the philosophy and ideas underpinning engagement in digital rather than real world learning.

1.4.6 Education in ICT

With regard to education of disabled people to enhance inclusion, Guo et al. (2005) found that disabled people with higher levels of education are more likely to have access to the Internet. Education of people with disabilities to access the Internet has received minimal coverage in the research literature (it is outside of the scope of this chapter to cover all of the literature on ICT education more generally). Despite some increase in recent years, there remain few studies reporting on programmes or interventions to improve the skills of different groups of people with disabilities to enhance the use of ITC.

Educational strategies need further development and exploration in the literature that focus on specific learning of groups who may be more systematically excluded, for example those with more severe ID and older disabled people. Education is likely to interact with age, with more educational and online opportunities being made available to younger people with disabilities than older people with disabilities (Parsons et al., 2008).

The focus on interventions for young people above lends support to this argument as there appears to be a disparity in how carers perceive the value of computers for adults and older adults with disabilities compared to children (Hoppestad, 2013). Moreover, older carers who are themselves less likely to access the Internet may be more likely to hold such attitudes (Parsons et al., 2008). Hence those with lifelong disabilities who are older and have yet to access the digital world need further consideration in practice and research as a group who are likely to experience even greater levels of digital exclusion.

Finally for this section, it is clear that people with impairments requiring greater levels of support have thus far been excluded from the online world and the digital revolution that has occurred (e.g. Gutiérrez & Martorell, 2011; Hoppestad, 2013). The ways in which ICT could benefit and support people with complex support needs is something which requires consideration in future research and practice so that the digital world does not simply become another space where such people are excluded.

1.5   Summary and conclusion

It is heartening to see a proliferation of studies into this area which have prized the voices of different groups with disabilities whose lives are most affected by these issues of digital inclusion and exclusion and that quantitative inferential research is tempered with more fine grain qualitative inquiries which include disabled voices and perspectives. There is scope for much more research into this area to better understand personal motivations, communication, negotiation of support and access, and how creative and positive risk-taking can utilised with regard to digital inclusion for people with disabilities.

Notably, the academic research material detailed herein is derived from a primarily global north perspective. Grech and Soldatic (2014) have highlighted the lack of consideration of the experiences and inclusion of people with disabilities in the global south, with much academic material remaining disengaged from Southern disability issues, with the applicability of theories, models, and findings from the global north being questioned. Hence, further research in digital inclusion and exclusion within these seldom considered contextually, culturally, economically, historically, geographically, and politically distinct regions is much needed.

To conclude, in tandem with a seemingly growing societal acceptance of disability, has been a proliferation of the digital world with ICT, the Internet and mobile and social networking technologies becoming ever more part of our everyday lives and an accompanying societal expectation of ability to engage with the digital world amongst its members. Due to rapid advancements, the world is developing into an increasingly complex place requiring greater skills to negotiate it effectively. Digital inclusion and exclusion are areas of contemporary relevance and concern, and researchers and practitioners in the field of cyberpsychology and more widely must take a nuanced approach to understand the many intrinsic and extrinsic facilitators and barriers to digital inclusion, particularly those pertaining to psychological, social, and cultural influences that drive the use of, and engagement with, online technology.

1.6   References

Alaszewski, A., & Alaszewski, H. (2002). Towards the creative management of risk: Perceptions, practices and policies. *British Journal of Learning Disabilities, 30*(2), 56–62.

Batorowicz, B., Campbell, F., von Tetzchner, S., King, G., & Missiuna, C. (2014) Social participation of school-aged children who use communication aids: The views of children and parents. *Augmentative and Alternative Communcation, 30*(3), 257–251.

Bourdieu, P. (1983). Forms of capital. In J. G. Richardson (Ed.), *Handbook of theory and research for the sociology of education* (pp. 241–258). New York: Greenwood Press.

Bowker, N., & Tuffin, K. (2003). Dicing with deception: People with disabilities' strategies for managing safety and identity online. *Journal of Computer-Mediated Communication, 8*(2). Retrieved from http://onlinelibrary.wiley.com/doi/10.1111/j.1083-6101.2003.tb00209.x/full

Brown, I. (2007). What is meant by intellectual and developmental disabilities? In I. Brown & M. Percy (Eds), *A comprehensive guide to intellectual and developmental disabilities* (pp. 3–16). Baltamore: Brookes.

Buell, S. (2015). Personal Communication.

Bunning, K., Heath, B., & Minnion, A. (2010). Interaction between teachers and students with intellectual disability during computer-based activities: The role of human mediation. *Disability & Technology, 22*(1–2), 61–71.

Castells, M. (1997). *The power of identity.* Malden, MA: Blackwell.

Caton, S., & Chapman, M. (In Press). The use of social media and people with intellectual disability: a systematic review and thematic analysis. *Journal of Intellectual & Developmental Disabilities.*

Chadwick, D. D., Fullwood, C., & Benton, C. (2015) An online life like any other: Social media experiences of people with intellectual disabilities. *13th Biannual Research Conference of the Nordic Network on Disability Research,* Bergen, Norway, 6–8 May, 2015.

Chadwick, D. D., Quinn, S., Fullwood, C., & Chen-Wilson, J. (2015). Public perceptions of the risks and benefits of being online for people with an intellectual disability. *13th Biannual Research Conference of the Nordic Network on Disability Research,* Bergen, Norway, 6–8 May, 2015.

Chadwick, D., Fullwood, C., & Wesson, C. J. (2013). Intellectual disability, identity and the Internet. In R. Luppicini (Ed.), *Handbook of research on techno-self: identity in a technological society* (pp. 229–254). USA: IGI Global.

Chadwick, D., Wesson, C. J., & Fullwood, C. (2013). Internet access by people with intellectual disabilities: Inequalities and ppportunities. *Future Internet, 5,* 376–397.

Cheatham, L. P. (2012). Effects of Internet use on well-being among adults with physical disabilities: A review. *Disability and Rehabilitation: Assistive Technology, 7*(3), 181–188.

Cohen, S., & Wills, T. A. (1985). Stress, social support, and the buffering hypothesis. *Psychological Bulletin, 98*(2), 310–357.

Coleman, J. S. (1988). Social capital in the creation of human capital. *American Journal of Sociology, 94 (Supplement)*, S95–S120.

Cromby, J., & Standen, P. (1999). Cyborgs and stigma: Technology, disability, subjectivity. In A. J. Gordo-Lopez & I. Parker (Eds.), *Cyberpsychology* (pp. 95–112). New York: Routledge.

Cromby, J., Standen, P., & Brown, D. J. (1996). The potentials of virtual environments in the education and training of people with learning disabilities. *Journal of Intellectual Disability Research,* 40(6), 489–501.

D'Aubin, A. (2007). Working for barrier removal in the ICT area: Creating a more accessible and inclusive Canada. *The Information Society, 23*(3), 193–201.

Dawe, M. (2006). Desperately seeking simplicity: How young adults with cognitive disabilities and their families adopt assistive technologies. *Proceedings of the CHI Conference,* Montreal, Quebec, Canada, 2006.

Didden, R., Scholte, R. H. J., Korzilius, H., de Moor, J. M. H., Vermeulen, A., O'Reilly, M., Lang, R., & Lancioni, G. E. (2009). Cyberbullying among students with intellectual and developmental disability in special education settings. *Developmental Neurorehabilitation, 12*(3), 146–151.

Dobransky, K., & Hargittai, E. (2006). The disability divide in internet access and use. *Information, Communication and Society, (Special Issue: Disability, Identity, and Interdependence: ICTs and New Social Forms), 9*(3), 313–334.

Drainoni, M. B., Houlihan, B., Williams, S., Vedrani, M., Esch, D., Lee-Hood, E., & Weiner, C. (2004). Patterns of internet use by persons with spinal cord injuries and relationship to health-related quality of life. *Archives of Physical and Medical Rehabilitation, 85*(11), 1872–1879.

Ellison N. B., Steinfield, C., & Lampe, C. (2007). The benefits of Facebook 'friends': Social capital and college students' use of the online network sites. *Journal of Computer-Mediated Communication, 12*(4), 1143–1168.

Feng, J., Lazar, J., Kumin, L., & Ozok, A. (2008). Computer usage by young individuals with down syndrome: An exploratory study. *Proceedings of Assets '08 Proceedings of the 10th International ACM SIGACCESS Conference on Computers and Accessibility,* Nova Scotia, Canada, 13–15 October 2008; 35–42.

Fichten, C. S., Ferraro, V., Asuncion, J. V., Chwojka, C., Barile, M., Nguyen, M. N., Klomp, R., & Wolforth, J. (2009). Disabilities and e-learning problems and solutions: An exploratory study. *Educational Technology & Society, 12*(4), 241–256.

Finkel, E. J., Eastwick, P. W., Karney, B. R., Reis, H. T., & Sprecher, S. (2012). Online dating: A critical analysis from the perspective of psychological science. *Psychological Science in the Public Interest, 13*(1), 3–66.

Foley, A., & Ferri. B. A. (2012). Technology for people, not disabilities: Ensuring access and inclusion. *Journal of Research in Special Educational Needs, 12*(4), 192–200.

Forrester-Jones, R., Carpenter, J., Coolen-Schrijner, P., Cambridge, P., Tate, A., Beecham, J., Hallam, A., Knapp, M., & Wooff, D. (2006). The social networks of people with intellectual disability living in the community 12 years after resettlement from long-stay hospitals. *Journal of Applied Research in Intellectual Disability,19*(4), 285–295.

Fullwood, C., & Wootton, N. (2009). Comforting communication in an online epilepsy forum. *Journal of Cybertherapy and Rehabilitation, 2*(2), 159–164.

Garcia, I. E., O'Brien, P., McConkey, R., Wolfe, M., & O'Doherty, S. (2014). Identifying the key concerns of Irish persons with intellectual disability. *Journal of Applied Research in Intellectual Disabilities, 27*(6), 564–575.

Gillespie-Lynch, K., Kapp, S. K., Shane-Simpson, C., Smith, D., & Hutman, T. (2014). Intersections between the autism spectrum and the Internet: Perceived benefits and preferred functions of computer-mediated communication. *Intellectual and Developmental Disabilities, 52*(6), 456–469.

Goggin, G., & Newell, C. (2003). *Digital disability: The social construction of disability in new media.* Lanham: Rowman & Littlefield.

Grech, S., & Soldatic, K. (2014). Introducing Disability and the Global South (DGS): We are critical, we are open access! *Disability and the Global South, 1*(1), 1–4.

Guo, B., Bricout, J. C., & Huang, J. (2005). A common open space or a digital divide? A social model perspective on the online disability community in China. *Disability & Society, 20*(1), 49–66.

Gutiérrez, P., & Martorell, A. (2011). People with intellectual disability and ICTs. *Communicar, 18*(36), 173–180.

Hatton, C., & Emerson, E. (1995). Staff in services for people with learning disabilities: An overview of current issues. *Mental Handicap Research, 8*(4), 215–219.

Helsper, E. (2008) *Digital inclusion: an analysis of social disadvantage and the information society.* Department for Communities and Local Government, London, UK.

Higher Education Statistics Agency (2015). *Higher Education Student Enrolments and Qualifications Obtained at Higher Education Providers in the United Kingdom 2013/14.* Retrieved from https://www.hesa.ac.uk/sfr210

Hollier, S. (2012). *Sociability and social media for people with a disability.* Report for Media Access Australia. 2012.

Holmes, K. M., & O'Laughlin, N. (2012). The experiences of people with learning disabilities on social networking sites. *British Journal of Learning Disabilities, 42*(1), 1–5.

Hoppestad, B. S. (2013). Current perspective regarding adults with intellectual and developmental disabilities accessing computer technology. *Disability & Rehabilitation: Assistive Technology, 8*(3), 190–194.

Hynan, A., Goldbart, J., & Murray, J. (2015a). A grounded theory of Internet and social media use by young people who use augmentative and alternative communication (AAC). *Disability & Rehabilitation,* early online.

Hynan, A., Murray, J., & Goldbart, J. (2015b). 'Happy and excited': Perceptions of using digital technology and social media by young people who use augmentative and alternative communication. *Child Language, Teaching & Therapy, 30*(2), 175–186.

Katz, E., Blumler, J., & Gurevitch, M. (1974). Utilization of mass communication by the individual. In J. Blumler & E. Katz (Eds.), *The uses of mass communication: Current perspectives on gratifications research* (pp. 19–34). Beverly Hills, CA: Sage.

Kent, M. (2015). Disability and eLearning: Opportunities and barriers. *Disability Studies Quarterly, 35*(1). Retreived from http://dsq-sds.org/article/view/3815

Kirk, S. (2008). Transitions in the lives of young people with complex healthcare needs. *Child: Care, Health & Development, 34*(5), 567–575.

Lathouwers, K., de Moor, J., & Didden, R. (2009). Access to and use of Internet by adolescents who have a physical disability: A comparative study. *Research in Developmental Disabilities, 30*(4), 702–711.

Livingstone, S., & Haddon, L. (2009). *EU Kids Online Final report.* LSE, London: EU Kids Online. (EC Safer Internet Plus Programme Deliverable D6.5)

Löfgren-Mårtenson, L. (2008). Love in Cyberspace: Swedish young people with intellectual disabilities and the internet. *Scandinavian Journal of Disability Research, 10*(2), 125–138.

McLaren, J., & Bryson, S. (1987). Review of recent epidemiological studies of mental retardation: Prevalence, associated disorders, and etiology. *American Journal of Mental Retardation, 92*(3), 243–254.

McNaughton, D., & Bryen, D.N. (2007). AAC technologies to enhance participation and access to meaningful societal roles for adolescents and adults with developmental disabilities who require AAC. *Augmentative & Alternative Communication, 23*(3), 217–229.

McVilly, K. R., Stancliffe, R. J., Parmenter, T. R., & Burton-Smith, R. M. (2006). 'I get by with a little help from my friends': Adults with intellectual disability discuss loneliness. *Journal of Applied Research in Intellectual Disabilities, 19*(2), 191–203.

Magee, J. J. & Betke, M. (2013). Automatically generating online social network messages to combat social isolation of people with disabilities. *Universal Access in Human-Computer Interaction, 8010*, 684–693.

Manion, M. & Bersani, H. (1987). Mental retardation as a western sociological construct: A cross-cultural analysis disability. *Handicap & Society, 2*(3), 231–245.

Morgan, S. (2004). Positive risk taking: An idea whose time has come. *Health Care Risk Report, 10*(10), 18–19.

Obst, P. & Stafurik, J. (2010). Online we are all able bodied: Online psychological sense of community and social support found through membership in disability-specific websites promotes well-being for people living with a physical disability. *Journal of Community and Applied Social Psychology, 20*525–531.

Office For National Statistics (ONS). (2013). *Internet Access Quarterly Update, Q4 2013*. Retrieved from http://www.ons.gov.uk/ons/rel/rdit2/internet-access-quarterly-update/2013-q4/stb-ia-q4-2013.html.

Oliver, M. (2013). The social model of disability: Thirty years on. *Disability & Society, 28*(7), 1024–1026.

Palmer, S. B., Wehmeyer, M. L., Davies, D. K., & Stock, S. E. (2012). Family members' reports of the technology use of family members with intellectual and developmental disabilities. *Journal of Intellectual Disability Research, 56*(4), 402–414.

Park, B. (1986). A method for studying the development of impressions of real people. *Journal of Personality and Social Psychology, 51*(5), 907–917.

Parsons, S., Beardon, L., Neale, H. R., Reynard, G., Eastgate, R., Wilson, J. R., Cobb, S. V. G., Benford, S. D., Mitchell, P., & Hopkins, E. (2000). Development of social skills amongst adults with Asperger's syndrome using virtual environments: The 'as interactive' project. *Proceedings of the 3rd International Conference of Disability, Virtual Reality and Associated Technologies*, Alghero, Italy: 163–170.

Parsons, S., Daniels, H., Porter, J., & Robertson, C. (2008). Resources, staff beliefs and organizational culture: Factors in the use of information and communication technology for adults with intellectual disabilities. *Journal of Applied Research in Intellectual Disability, 21*(1), 19–33.

Perske, R. (1972). The dignity of risk. In W. Wolfensberger, B. Nirje, S. Olansky, R. Perske, & P. Roos (Eds.), *The principles of normalization in human services* (pp. 194–200). Toronto: National Institute on Mental Retardation through Leonard Crainford.

Plichta, P. (2011). Ways of ICT usage among mildly intellectually disabled adolescents. In E. Dunkels., G. M. Franberg, & C. Hallgren (Eds.), *Youth culture and net culture: Online social practices* (pp. 296–315). Hershey, PA: IGI Publishing.

Pueschel, S. M., & Scola, P. S. (1988). Parents' perception of social and sexual functions in adolescents with Down's syndrome. *Journal of Mental Deficiency Research, 32*(3), 215–220.

Putnam, R. D., & Goss, K. A. (2002). Introduction. In R. D. Putnam (Ed.), *Democracies in flux: The evolution of social capital in contemporary society* (pp. 3–19). New York: Oxford University Press.

Raghavendra, P., Wood, D., Newman, L., Lawry, J., & Sellwood, D. (2012). Why aren't you on Facebook? Patterns and experiences of using the Internet among youth with physical disabilities. *Technology & Disability, 24*(2), 149–162.

Raghavendra, P., Grace, E., Newman, L., Wood, D., & Connell, T. (2013). 'They think I'm really cool and nice': The impact of Internet support on the social networks and loneliness of young people with disabilities. *Telecommunications Journal of Australia, 63*(2), 22.1–22.15.

Ritchie, H., & Blank, P. (2003). The promise of the Internet for disability: A study of on-line services and web site accessibility at centers for independent living. *Behavioral Sciences & the Law, 21*(1), 5–26.

Roberts, J. B., Crittenden, L. A., & Crittenden, J. C. (2011). Students with disabilities and online learning: A cross-institutional study of perceived satisfaction with accessibility compliance and services. *Internet and Higher Education, 14*(4), 242–250.

Ruggerio, T. E. (2000). Uses and gratifications theory in the 21st century. *Mass Communication and Society, 3*(1), 3–37.

Schalock, R. L., Borthwick-Duffy, S. A., Bradley, M. et al. (2010). *Intellectual disability: definition, classification, and systems of supports* (11th ed). Washington: American Association on Intellectual and Developmental Disabilities.

Seale, J. (2003). Researching home page authorship of adults with learning disabilities: Issues and dilemmas. *Proceedings of International Education Research Conference AARE – NZARE*, Auckland, New Zealand, 30 November–3 December 2003.

Seale, J. (2014). The role of supporters in facilitating the use of technologies by adolescents and adults with learning disabilities: a place for positive risk-taking? *European Journal of Special Needs Education, 29*(2), 220–236.

Seale, J., Nind, M., & Simmons, B. (2012). Transforming positive risk-taking practices: The possibilities of creativity and resilience in learning disability contexts, *Scandinavian Journal of Disability Research, 15*(3), 1–16.

Sevens, L., Schuurman, I., Vandeghinste, V., & Van Eynde, F. (2015). Automatic translation with pictographs to serve people with IDD. *13th Biannual Research Conference of the Nordic Network on Disability Research,* Bergen, Norway, 6–8 May, 2015.

Shakespeare, T. (2014). *Disability rights and wrongs revisited.* London, New York: Routledge.

Shpigelman, C. N., & Gill, C. J. (2013). The characteristics of unsuccessful e-mentoring relationships for people with disabilities. *Qualitative Health Research, 23*(4), 463–475.

Shpigelman, C. N., Weiss, P. L., & Reiter, S. (2009). E-mentoring for all. *Computers in Human Behavior, 25*(4), 919–928.

Soderstrom, S. (2009). Offline social ties and online use of computers: A study of disabled youth and their use of ICT advances. *New Media & Society, 11*(5), 709–727.

Stendal, K. (2012). How do people with disability use and experience virtual worlds and ICT: A literature review. *Journal of Virtual Worlds Research, 5*(1), 1–17.

Stendal, K., Molka-Danielson, J., Munkvold, B., E., & Balandin, S. (2013). Social affordances for people with lifelong disability through using virtual worlds. *46th Hawaii International Conference on System Sciences.*

Sundqvist, A., & Ronnberg, J. (2010). A qualitative analysis of email interactions of children who use augmentative and alternative communication. *Augmentative & Alternative Communication, 26,* 255–266.

Third, A., & Richardson, I. (2009). *Analysing the impacts of social networking for young people living with chronic illness, a serious condition or a disability: An evaluation of the Livewire on-line community.* Retrieved from http://www.livewire.org.au/ resources/DOCUMENT/180510120528 Livewire Final% 20Report WEB.pdf.

Thoreau, E. (2006). Ouch!: An examination of the self-representation of disabled people on the Internet. *Journal of Computer-Mediated Communication, 11*(2), article 3.

Tilly, L. (2015). Able to Include; The role of technology in enabling inclusion and citizenship. *13th Biannual Research Conference of the Nordic Network on Disability Research,* Bergen, Norway, 6–8 May, 2015.

van Deursen, A. J. & van Dijk, J. A. (2014). The digital divide shifts to differences in usage. *New Media & Society. 16*(3), 507–526.

Valenzuela. S., Park, N., & Kee, K. F. (2009). Is there social capital in a social network site?: Facebook use and college students' life satisfaction, trust, and participation. *Journal of Computer-Mediated Communication, 14*(4), 875–901.

Vallor, S. (2011). Flourishing on Facebook: Virtue friendship and new social media. *Ethics & Information Technology, 14*(3), 185–199.

Vanden Abeele, M., de Cock, R., & Roe, K. (2012). Blind faith in the web? Internet use and empowerment among visually and hearing impaired adults: A qualitative study of benefits and barriers. *Communications, 37*(2), 129–151.

Warschauer, M. (2003). *Technology and social inclusion: Rethinking the digital divide.* Cambridge, MA: MIT Press.

Wehmeyer, M. L., Smith, S. J., Palmer, S. B., & Davies, D. K. (2004). The effect of student directed transition planning with a computer-based reading support program on the self-determination of students with disabilities. *Journal of Special Education Technology, 19*(4), 7–22.

Williams, D. (2006). On and off the 'net': Scales for social capital in an online era. *Journal of Computer-Mediated Communication, 11*(2), 593–628.

2
Parenting the Online Child

Sally Quinn

2.1 Introduction

Children are now living in a multi-media world (Livingstone, 2007) where the online world in particular has its own unique set of risks, such as cyberbullying, grooming, and invasion of privacy (Hasebrink et al., 2009). These online risks bring new parental concerns. Consequently, there has been a host of research which has focused on parental mediation strategies aimed at enhancing children's experiences of the online world while minimising potential risks. For example, Duerager and Livingstone (2012) show that various different strategies are used by parents across Europe to promote safe use of the Internet. For example, 96% of parents of 9 to 12 years olds report that they do not allow their child to give out personal information to others online and 85% talk to their child about what they do online. It is therefore important to understand what strategies parents are using, what other factors might be related to the degree of use of these strategies and how effective these strategies are at optimising children's online experiences. This chapter will first discuss theoretical perspectives on categorising parental mediation of children's Internet use, looking at two different approaches. The focus will then turn to factors which have been found to be associated with the levels and types of strategies used by different parents. The chapter will end by considering how theory and research evidence may be applied to effectively reduce negative online experiences and increase positive aspects of being online. The content of this chapter will focus on parental mediation strategies used among children in middle childhood and adolescence (approximately ages 9 to 18), since this is where the majority of research has been done.

2.2 Theoretical approach to online parenting

Before the various theories on online parenting are discussed, it is perhaps important to consider the risks children are experiencing online, as well as the types of opportunities they are exposed to. There are various forms of risk that children can encounter while online and some may be more prevalent than others but these prevalence rates may change over time. Livingstone et al. (2014) report on the prevalence rates from two different projects with children from several European countries. The EU Kids Online project provided data from 2010 and the Net Children Go Online project provided data from 2014. Comparisons from these studies showed that there were moderate increases in some risks. Specifically, an increase in cyberbullying was found (8% to 12%) and a small increase in the percentage of children who reported seeing sexual images online (18% to 20%). Seeing websites with inappropriate content (e.g. hate messages, drugs, self-harm, and the promotion of eating disorders) also increased between 2010 and 2014, although only moderately. Some risks were seen to decrease. Namely, receiving sexual messages (14% to 12%) and making contact with someone online who they have never met offline (32% to 29%). Livingstone et al. (2014) report that the increases in some risks may go hand in hand with increased use of the Internet. This is supported by their data which also showed an increase in the different activities children engage in while online. The most common activities were visiting social networking sites (which increased from 44% to 63%) and watching videos on sites such as YouTube (which increased from 32% to 59%). Increases were also seen in using instant messaging platforms, posting or looking at videos or photographs, playing games, and downloading music.

Since the Internet clearly has both risks and opportunities associated with its use by children, parents play a crucial role in how children experience the online world. It is therefore important to explore the different types of strategies parents use and how these strategies have informed the development of different theoretical taxonomies relating to parenting the online child. Parental mediation of children's use of media is not a new topic. For example, research has identified strategies parents use to mediate their children's use of television. Within this context, Nathanson (1999, 2001) outlines three main parental mediation strategies: (i) Restrictive (rules about the use of media), (ii) Active (supporting the child's use of media), and (iii) Co-use (sharing the child's use of media). These have since been modified to focus on parental mediation of the Internet specifically.

Building on Nathanson's (1999, 2001) approach, Livingstone and Helsper (2008) carried out an analysis on parental mediation strategies relating to children's Internet use with a sample of parents of 9 to 17 year olds. They demonstrated that the strategies outlined by Nathanson in terms of pre-Internet media have been extended to Internet use but with slight modifications and additions. Their analysis identified four different types of strategy: (i) *"Active co-use"* refers to strategies such as parents sitting with their child while they are online or helping their child with any online issues they may experience. This strategy clearly reflects parental involvement in children's online experiences by sharing online experiences with them. (ii) *"Technical restrictions"* refer to strategies whereby software is installed to monitor and filter online content. This is a way of restricting children's Internet use without direct involvement from the parent; the software does this for them. (iii) *"Interactive restrictions"* refer to those rules which govern when the child can go online or the kind of activities the child is allowed to do online. This type of restrictive strategy requires direct involvement from the parent; parents must tell the child what the rules are surrounding their use of the Internet. (iv) *"Monitoring"* refers to parental checks of what the child has been doing online (e.g. checking content of emails). These four different types of parental mediation have formed the basis of much of the research into how children are parented while online (e.g. Duerager & Livingstone, 2012; Kirwil, 2009; Lee, 2012; Nikken & Jansz, 2014; Sonck et al., 2011, 2013). Livingstone and Helsper (2008) discuss how these modifications and additions to the original parental mediation strategies used for television viewing have emerged due to the interactive nature of the Internet. Parents have needed to develop their strategies to encompass such things as the type of information children can access and with whom they may be able to interact. Issues that were perhaps not applicable to pre-Internet media use such as television viewing.

In addition to this categorisation of parental mediation, there has also been literature which links Internet parenting to *general* parenting styles found in research from Developmental Psychology. The theoretical basis of parenting styles outlined by Baumrind (1966) are now widely accepted in psychological research, namely Authoritative, Authoritarian, and Permissive and this was further developed by Maccoby and Martin (1983) to include a fourth parenting style, "Uninvolved" or "Neglectful." Each of these styles are characterised by varying levels of "control" and "warmth." Control refers to the enforcement of rules, whereas Warmth refers to a more discursive approach to parenting, discussing rules with the child and allowing the child some level of autonomy (Baumrind, 1966). Authoritative parents use high levels of both control and warmth; Authoritarian parents use high levels of control but low levels of warmth; Permissive parents use low levels of control but high levels of warmth; and Neglectful parents use low levels of both control and warmth.

These parenting styles set out by Baumrind (1966) and Maccoby and Martin (1983) have also provided a theoretical framework for parental mediation of children's Internet use. Control strategies refer to strategies such as placing rules on when and where the child can go online, whereas Warmth strategies refer to elements of parenting such as talking to the child about what they have been doing online and sharing in their online experiences. These Control and Warmth strategies are used to varying degrees among different parents, with parents being categorised as using a particular parenting style (e.g. Authoritative) dependent on the degree of use of these strategies. Some research with the general population has found that the Authoritative style of parenting seems to be a popular style where parents use high levels of both control and warmth strategies (e.g. Leung, 2007; Lou et al., 2010; Valcke et al., 2011).

Although there are two different sets of parenting taxonomies emerging from different research backgrounds, it is clear that there are similarities between the two. That is, both have identified that parents might use different rules and regulations such as how long children are allowed online and where they can go online (e.g. bedroom or a less private space in the home). Both consist of strategies such as supervising children's Internet use through, for example, either being in close proximity to the child while they are online or watching what the child does online. In addition, both taxonomies have identified that parents may share their child's online experiences as a way of "keeping in the loop" about what their child is doing and that parents might talk to children about their experiences in order to keep the lines of communication open. Technology such as monitoring or blocking software may also be used. Although the literature seems to indicate that many parents might use high levels of both restrictive/control strategies and active/warmth strategies, there are factors which have been shown to be associated with the degree to which these mediation strategies are used by parents.

2.3 Factors associated with levels of parenting strategies

There is evidence to suggest that strategies may vary when child and/or parent characteristics are taken into account. First, the age of the child seems to play a role. It has been demonstrated that parents of younger children tend to implement more rules and set more limits compared to parents of older teenagers (Lee, 2012; Livingstone & Helsper, 2008; Wang et al., 2005). This is supported in a study of social media use where those teenagers who were found to have parents who adopted a Neglectful parenting style (low levels of both control and warmth) were older than those whose parents who adopted an Authoritarian or Authoritative style (Rosen et al., 2008). In addition, Helsper et al. (2013) report that children of parents who used higher levels of restrictive strategies were younger (aged 11 or 12 years) than those who used low levels of restrictive strategies (aged 14 years). In terms of online parenting, parents might acknowledge that with age, their child may be acquiring the skills they need in order to protect themselves from potential risk and harm and to use the Internet safely.

However, this linear relationship between parental strategies and age has not always been found. Kirwil et al. (2009) report that 11 to 14 year old children had higher restrictions placed on their Internet use than younger and older children. It is thought that this may be related to their increased use of the social side of the Internet at this age (e.g. using social networking sites such as Facebook) but still not having the necessary skills to protect themselves from the potential risks of using the Internet in this way (Livingstone & Bober, 2004). This study shows that the idea of a linear relationship between levels of parenting and age may be too simplistic. The way in which Internet use changes as children grow older may be a factor which contributes towards the levels of strategies chosen by parents. Helsper et al. (2013) propose that parents could increase their use of active strategies the more their child becomes experienced in navigating the online world and/or potentially experiences a greater level of risk. However, there is a lack of evidence to support this idea that the changes in Internet use *cause* changes in parental mediation. To date, it is unknown whether levels of parental strategies affect the aspects of the Internet the child uses and/or vice versa.

The age of the parent may also play a role in the level of strategies used. Valcke et al. (2010) found that parents between the ages of 45 and 54 years were more likely to use lower levels of both control and warmth than parents between the ages of 25 and 44 years. In addition, older parents are less likely to check which websites their child has been visiting (Wang et al., 2005). This could potentially be related to the level of exposure to the Internet. Younger parents have arguably lived with the Internet for a greater proportion of their lives and may therefore feel more comfortable parenting their child with this relatively new form of media. In addition to age, the education level of the parent has also been found to be associated with differing levels of strategies. Those who are educated to a higher level have been found to employ higher levels of control and warmth strategies (Valcke et al., 2010), but are less likely to use monitoring software (Wang et al., 2005). However, this positive relationship between education level and use of restrictive strategies has not always been supported. Helsper et al. (2013) report that parents who were educated to a lower level tended to employ very high levels of restrictive strategies; those who had the highest level of education were most likely to use active strategies. Pasquier et al. (2012) supported this, finding that highly educated parents were less likely to use restrictive strategies. Moreover, they also found that parents who were educated to a higher level were more likely to offer advice to children on how to use the Internet safely (an active strategy) compared to parents who were educated to a lower level. They attribute this to the likelihood that parents who are educated to a higher level may be more likely to be competent users of the Internet due to greater experience with the online world. Internet experience may also therefore play a role in the degree and type of strategies used by different parents.

There is some evidence to suggest that parental Internet use may be linked to parental mediation. In a study by Wang et al. (2005), parents who used the Internet were more likely to use filtering and monitoring software and to check the websites their child has visited compared to parents who do not use the Internet. However, this study did not examine frequency of Internet use by parents but rather whether the Internet was used by the parent or not. There may be great variation in the degree to which the Internet is used and thus the frequency of Internet use may provide a more accurate reflection of internet use. For example, parents with less Internet experience or those who consider themselves beginners are more likely to employ lower levels of both control and warmth strategies (Valcke et al., 2010) and those parents who see themselves as more skilled with the Internet and to use the Internet frequently are more likely to use all types of mediation (Livingstone & Helsper, 2008). These studies would seem to suggest that the increased skill or experience of the online environment acquired by parents may provide increased skills or confidence in managing children's online experiences. It may also make parents more aware of the potential risks the Internet can pose to their child and therefore react by putting in place higher levels of strategies aimed at protecting their child. Nevertheless, Kirwil et al. (2009) report that there is a lack of clarity on the relationship between parental Internet experience and activities that children are allowed to take part in online. From their data (taken from the EU Kids

Online Project), they show only small differences in restrictions on child online activities between parents who use the Internet regularly and those who do not – those who used the Internet daily and those who used the Internet rarely or never imposed the lowest level of restrictions (but only slightly).

Hence, the links between parental Internet use and the mediation strategies employed are not clear. Some research shows clear links between these two factors while others show only weak links. Parental Internet use is often taken as a proxy for measuring parental Internet skill (Kirwil et al., 2009), but it may be important to consider that the *type* of activities parents engage in online as a better predictor of Internet skill than frequency of use. There is little research which investigates this link. Among a sample of Korean children and their parents, Lee (2012) measured parental skills by asking parents about their ability to do 10 online activities (e.g. playing games, online shopping, being involved in an online community). The results showed that parental online skills positively predicted the degree of restrictive strategies used by parents. This evidence supports the majority of the research discussed above and demonstrates that when Internet skill is measured by the types of activities done online rather than just the frequency of Internet use, parents with greater skill are more likely to implement higher levels of restrictive strategies.

There have, however, been mixed findings with regard to some factors which might be associated with Internet parenting. For example, parents who hold a positive attitude about the Internet have been found to use high levels of control (Valcke et al., 2010) but other studies have found that high levels of all parental mediation are related to more concern about online risks (Nikken & Jansz, 2011). This inconsistency is confounded further by research which has found no link between attitudes towards the Internet and Internet parenting strategies (e.g. Wang et al., 2005). In addition to attitudes towards the Internet, the gender of the parent has also received mixed results. Mothers have been found to use higher levels of both control and warmth compared to fathers (Valcke et al., 2010) and to play a more active role in parenting their child while they are online (Kirwil et al., 2009). This is supported by Livingstone (2007) who reports that children view their mothers as more restrictive than their fathers. However, other studies have found no differences between mothers and fathers in terms of rule setting or using monitoring software (Wang et al., 2005). The evidence which suggests a relationship between these factors and Internet parenting strategies is therefore less convincing.

There is also evidence that there are differences between countries in the way children are parented online, at least in Europe. The most recent data to emerge in terms of parental mediation of the Internet comes from the EU Kids Online project. The EU Kids Online project has carried out extensive research into parental mediation of the Internet across different European countries. In one report stemming from this project, Helsper et al. (2013) show four types of parenting styles: (i) Restrictive mediation (parents who use high levels of restrictive strategies and a moderate amount of active mediation strategies); (ii) Active mediation (parents who use high levels of active mediation and moderate levels of restrictive mediation strategies); (iii) Passive mediation (parents use relatively low levels of all mediation, in particular active mediation, and monitoring and restrictive strategies); (iv) All-rounders (parents use high levels of restrictive, active, and monitoring and technical restrictive strategies). Although they found evidence of every style in all 25 countries involved in the research, each country was found to favour one of the four parenting styles such that countries would cluster together. For example, Austria, Belgium, France, Germany, Greece, Ireland, Italy, Portugal, Spain, Turkey, and the UK clustered together around the Restrictive mediation style while Scandinavian countries (Denmark, Finland, Netherlands, Norway, and Sweden) clustered together around the Active mediation style. This report shows clear evidence that there are potential cultural, social, and economic differences which influence parenting styles dependent on the country of residence (Helsper et al.). However, it is beyond the scope of this chapter to discuss these cultural differences in any further depth. Readers may however refer to Chapter 3 for a broader discussion of the manner in which culture impacts upon online engagement.

This body of literature has demonstrated that there may be factors associated with the levels of strategies that parents use when managing their child's online activities. However, due to the lack of longitudinal data, it is difficult to establish the direction of causation with many of these factors. Nevertheless, it is important to attempt to gain some understanding of the effectiveness of parental mediation strategies at optimising children's online experiences.

2.4 Effectiveness of parenting strategies

There has been much support for parents who show a high degree of control and warmth strategies, or who have been categorised as Authoritative. Higher levels of restrictive and active strategies have been found to be related to less exposure to online risk and to lower levels of harm among some children (Duerager & Livingstone, 2012). Further evidence of this has been found from research which has looked at specific types of risk. Children of Authoritative parents are less likely to meet an online only friend offline (Rosen et al., 2008) and children of parents who use high levels of warmth and control strategies are less likely to disclose personal information online (Lwin et al., 2008). With regard to cyberbullying, Mesch (2009) reports that children are less likely to be bullied when limitations are placed on the websites they are allowed to visit and when parents monitor the websites visited. In addition, there has been some evidence that enforcing rules about content (a control strategy) as well as having high quality communication between parent and child (a warmth strategy) can reduce the likelihood of compulsive Internet use (van den Eijnden et al., 2010). In terms of children using the social side of the Internet, Livingstone and Helsper (2008) report that interactive restrictions (e.g. the child being prohibited to use email, chat rooms, or instant messaging services) were beneficial in reducing overall online risks. There is therefore some support for using high levels of both active/warmth and restrictive/control strategies, particularly for some types of risk.

However, high levels of warmth and control strategies have not always been found to protect children from some risks. Livingstone and Helsper (2008) report that active co-use and using filtering and monitoring software were ineffective at reducing online risks such as exposure to pornography or violent material, giving out personal information and meeting an online friend offline. In terms of restrictive strategies specifically, the enforcement of Internet rules has been reported to actually increase the risk of addictive use of the Internet (Lee, 2012) and restrictions on time, rules about sharing information online and the location of the home computer have been found to have no effect on the risk of cyberbullying (Mesch, 2009). Duerager and Livingstone (2012) report that technical strategies (such as using blocking or monitoring software) had no effect on levels of exposure to risk and only had an effect on reported levels of harm for older children (15/16 year olds). Similar results have been found in terms of the potential ineffectiveness of active strategies. The frequency of communication between parent and child has been found to have no relationship with compulsive Internet use (van den Eijnden et al., 2010). Moreover, Duerager and Livingstone (2012) also report that active strategies related to safety (e.g. talking to the child about what to do if they experience something online which bothers them) were related to an increase in reported levels of risk and harm.

There is also research to indicate that even though some strategies may be effective to some degree, they explain very little of the negative outcome. Valcke et al. (2011) found that although higher control strategies were related to lower risks such as meeting up with someone who the child met online, this relationships was very weak meaning that other factors play a role in explaining why children take part in this risky behaviour. This suggests that there are many other factors influencing children's exposure to risk online, other than the level of parenting strategies. There may also be differences between countries with different child-rearing cultures. For example, Kirwil (2009) distinguishes between countries which favour individualistic values (i.e. those who favour tolerance, respect, responsibility, independence, imagination, and lack of hard work) and those who favour collectivistic values (i.e. obedience, thrift, religious faith, determination, and good manners), reporting that placing time limits on children's use of the Internet may have different effects in these cultures. In individualistic cultures (e.g. Belgium, Denmark, Ireland, the Netherlands, Sweden, and the UK) Kirwil found that this strategy was related to an increase in children's experiences of online risk whereas in collectivistic cultures (e.g. Bulgaria, Czech Republic, Estonia, Poland, and Portugal) this strategy was related to a decrease in exposure to online risks. Hence, aspects associated with one's environment (e.g. culture, social, and economic factors) might influence the effectiveness of certain strategies.

In their report for the EU Kids Online project, Helsper et al. (2013) identified that the countries which fell into the Restrictive cluster were also the countries with the highest percentage of children reporting no experience of risk while online. The results from this report would therefore suggest that using high levels of restrictive strategies protects children from experiencing online risk. However, experiencing risk is not the same as experiencing harm. Livingstone et al. (2011) report that in their European sample, 14% of nine to 16 year olds reported seeing a sexual image online but 72% of these reported that they were not upset at all by being exposed to the image. There is therefore a degree of resilience among some children meaning that exposure to risk does not inevitably lead to harm. There is some suggestion that those children who have greater resilience are those who are exposed to risk more often (Duerager & Livingstone, 2012) but clearly, the probability of exposure to risk needs to be managed in line with the child's development. From this research, it would therefore seem that although restrictive/control strategies leads to less exposure to risk, this might not be a wholly positive thing; children need a degree of exposure to risk in order to develop resilience and coping skills.

The strategies and styles that parents use may therefore protect against some risks but in some circumstances they may not be providing the protection parents desire. There may also be important differences between countries and cultures in terms of the effectiveness of strategies in protecting children from risk. In addition, as already discussed in this chapter, there are several factors influencing the strategies parents use (e.g. age of the child, parental Internet experience) but there is little research which can show the interaction effect of these factors with the parenting strategies used and how they affect the child's Internet experience. Moreover, it is unclear whether parents are using strategies proactively (i.e. to prevent children experiencing risk) or reactively (i.e. in response to children already experiencing some risk and to prevent them experiencing more). Duerager and Livingstone (2012) speculate that strategies such as restrictive and active use strategies which relate to use of the Internet are used proactively to try to diminish exposure to risk and harm, whereas monitoring and active strategies relating to safety (e.g. suggesting ways to use the Internet safely) are used reactively. That is, these latter strategies are used in response to some negative event already experienced online. This is suggested in light of their data which shows that higher levels of these strategies are related to more risk and harm rather than less.

The Internet undoubtedly has benefits for children and teenagers of today's society (Tynes, 2007) and it is important for parents to consider this when setting levels of parenting. However, few studies have examined the effects of the different parenting styles and mediation strategies on these potential positive benefits. Warmth strategies have been found to be positively related to using the Internet for educational activities and to increase family communication (Lee & Chae, 2007). However, it has been suggested that filtering software (a control strategy) may block important educational websites (Canyaka & Odabasi, 2009) and whilst employing interactive restrictions (such as limiting communication online via emails, chat rooms, and instant messaging services) may reduce online risks, it may also restrict the benefits of using the Internet to interact with others (Livingstone & Helsper, 2008). There is some research which points to a positive relationship between communicating online with friends and friendship effects such as feelings of belonging (Quinn & Oldmeadow, 2013) and closeness to friends (Valkenburg & Peter, 2007). In addition, although Helsper et al. (2013) report that higher restrictive strategies seem to help reduce risk, they also report that this strategy might restrict access to opportunities; children whose parents use high levels of restrictive strategies are less likely to experience the opportunities and benefits of the Internet. For example, specifically within the UK, fewer children are experiencing positive aspects such as networking, gaming and exploring the online world (Helsper et al., 2013). The authors acknowledge that this strategy might not therefore be the most effective when attempting to optimise children's Internet experiences since it offers protection at the cost of exposing children to the benefits that the Internet has to offer.

The research which attempts to gauge the effectiveness of parental mediation of children's Internet use demonstrates that this issue is complex. There is no evidence that "one-size-fits-all," nor is there evidence that one particular parenting style leads to lower risk and harm and to a greater exposure to opportunities and benefits online (Helsper et al., 2013). Although some strategies may be effective at reducing some element of risk for some, this may be done at the cost of children missing out on the potential benefits of being online or at the cost of stunting their development of resilience. Moreover, the characteristics of parents and children (e.g. age, Internet experience) undoubtedly has an impact on which strategies parents choose to use and at what level. Nevertheless, there are many agencies which offer help and advice to parents on how to mediate their child's Internet use. For example, in the UK, the Child Exploitation and Online Protection Agency (CEOP) together with the National Crime Agency (NCA) provide parents with advice on how to help their children make the most of being online (CEOP, 2012). One of the key pieces of advice to parents from this initiative is that parents should try to get involved with their child's online activities by, for example, talking to them about the sites they visit and coming to a mutual agreement about Internet rules. This advice fits with the active co-use/warmth strategies categorisation of parenting strategies and is also echoed by key stakeholders (Helsper et al., 2013). These types of strategies enable children to be able to talk to their parents should they experience any risk online and there is evidence that these particular strategies are not only related to reduced online risk but also reduced harm from exposure to this risk (Duerager & Livingstone, 2012). Since there is also evidence that parents with less Internet experience may be less likely to use these types of strategies, schools and other organisations could provide support for these parents by giving guidance on how to implement these more active types of strategies and/or offering them opportunities to become more "Internet savvy."

2.5 Summary

The identification of the different strategies used by parents to mediate their child's Internet use has shown that parents may use rules surrounding the use of the Internet, monitoring and blocking software, as well as strategies which involve sharing and talking to children about their online experiences. Although there is evidence to suggest that some of these strategies are effective at reducing risk and harm while online, it is important to also consider the effects of exposure to online opportunities and benefits. In addition, there are factors associated with the characteristics of child and parent as well as potential factors associated with different countries which have been found to be associated with the use of different strategies. However, there is little research which takes all these factors into account when examining the effectiveness of these strategies. In addition there is a lack of longitudinal evidence to help parents understand the long-term effects of the different parenting strategies as well as the long term effects of exposure to risk and harm. Until there is evidence to indicate which strategies are most effective for which groups of parents and children and the potential long term effects of these strategies (if this can ever be achieved), one of the main pieces of current advice is for parents to stay involved in their child's use of the Internet and to keep the lines of communication open. Parents will then be more likely to understand how their child is using the Internet and, perhaps more importantly, children are more likely to feel they have a trusted adult to whom they can turn should they experience risk (or indeed harm) online.

2.6   References

Baumrind, D. (1966). Effects of authoritative parental control on child behavior. *Child Development, 37*(4), 887–907.

Canyaka, S., & Odabasi, H. F. (2009). Parental controls on children's computer and internet use. *World Conference of Educational Studies – New Trends and Issues in Educational Sciences, 1*(1), 1105–1109.

CEOP. (2012). The parents' and carers' guide to the internet. Retrieved from https//www.thinkuknow.co.uk/parents/parentsguide/

Duerager, A., & Livingstone, S. (2012). *How can parents support children's internet safety?* EU Kids Online, London, UK.

Hasebrink, U., Livingstone, S., Haddon, L., & Olafsson, K. (2009). *Comparing children's online opportunities and risks across Europe: Cross-national comparisons for EU Kids Online*. London: EU Kids Online (Deliverable D3.2, 2nd edition).

Helsper, E. J., Kalmus, V., Hasebrink, U., Sagvari, B., & de Haan, J. (2013). *Country classification: Opportunities, risks, harm and parental mediation*. London: EU Kids Online.

Kirwil, L. (2009). Parental mediation of children's internet use in different European countries. *Journal of Children and Media, 3*(4), 394–409.

Kirwil, L., Garmendia, M., Garitaonandia, C., & Fernandez, G. M. (2009). Parental mediation. In S. Livingstone & L. Haddon (Eds.), *Kids online. Opportunities and risks for children* (pp. 199–215). Bristol: The Policy Press.

Lee, S.-J. (2012). Parental restrictive mediation of children's internet use: Effective for what and for whom? *New Media & Society, 15*(4), 466–481.

Lee, S. & Chae, Y. (2007). Children's internet use in a family context: Influence on family relationships and parental mediation. *Cyberpsychology and Behavior,* 10(5), 640–644.

Leung, L. (2007). Stressful life events, motives for Internet use, and social support among digital kids. *Cyberpsychology & Behavior, 10*(2), 204–214.

Livingstone, S. (2007). Strategies of parental regulation in the media-rich home. *Computers in Human Behavior, 23*(2), 920–941.

Livingstone, S., & Bober, M. (2004). *UK children go online: Surveying the experiences of young people and their parents.* London: London School of Economics and Political Science. Retrieved February 4, 2011 from http://eprints.lse.ac.uk/395/1/UKCGOsurveyreport.pdf.

Livingstone, S., Haddon, L., Gorzig, A., & Olafsson, K. (2011). *Risks and safety on the internet: The perspective of European children.* Full Findings. LSE, London: EU Kids Online.

Livingstone, S., & Helsper, E. J. (2008). Parental mediation of children's internet use. *Journal of Broadcasting & Electronic Media, 52*(4), 581–599.

Livingstone, S., Mascheroni, G., Ólafsson, K., & Haddon, L. (2014) *Children's online risks and opportunities: Comparative findings from EU Kids Online and Net Children Go Mobile.* EU Kids Online, LSE: London, UK.

Lou, S. J., Shih, R. C., Liu, H. T., Guo, Y. C., & Tseng, K. H. (2010). The influences of the sixth graders' parents' internet literacy and parenting style on internet parenting. *Turkish Online Journal of Educational Technology, 9*(4), 173–184.

Lwin, M. O., Stanaland, A. J. S., & Miyazaki, A. D. (2008). Protecting children's privacy online: How parental mediation strategies affect website safeguard effectiveness. *Journal of Retailing, 84*(2), 205–217.

Maccoby, E. E., & Martin, J. A. (1983). Socialization in the context of the family: Parent-child interaction. In P. H. Mussen & E. M. Hetherington (Eds.), *Handbook of child psychology: Vol 4.socialization, personality and social development* (Vol. 4, pp. 1–101). New York: Wiley.

Mesch, G. S. (2009). Parental mediation, online activities, and cyberbullying. *Cyberpsychology & Behavior, 12*(4), 387–393.

Nathanson, A. I. (1999). Identifying and explaining the relationship between parental mediation and children's aggression. *Communication Research, 26*(2), 124–143.

Nathanson, A. I. (2001). Parent and child perspectives on the presence and meaning of parental television mediation. *Journal of Broadcasting & Electronic Media, 45*(2), 201–220.

Nikken, P., & Jansz, J. (2011). Parental mediation of young children's internet use. *EU Kids Online II Conference, London.* London.

Nikken, P., & Jansz, J. (2014). Developing scales to measure parental mediation of young children's internet use. *Learning, Media and Technology, 39*(2), 250–266.

Pasquier, D., Simoes, A., & Kredens, E. (2012). Agents of mediation and sources of safety awareness: A comparative overview. In S. Livingstone, L. Haddon & A. Görzig (Eds.), *Children, risk and safety on the internet. Research and policy challenges in comparative perspective* (pp. 219–230). Bristol: The Policy Press.

Quinn, S., & Oldmeadow, J. A. (2013). Is the igeneration a 'we' generation? Social networking use among 9- to 13-year-olds and belonging. *The British Journal of Developmental Psychology, 31*(Pt 1), 136–142.

Rosen, L. D., Cheever, N. A., & Carrier, L. M. (2008). The association of parenting style and child age with parental limit setting and adolescent MySpace behavior. *Journal of Applied Developmental Psychology, 29*(6), 459–471.

Sonck, N., Livingstone, S., Kuiper, E., & de Haan, J. (2011). *Digital literacy and safety skills*. London, UK: EU Kids Online, London School of Economics & Political Science.

Sonck, N., Nikken, P., & de Haan, J. (2013). Determinants of internet mediation. *Journal of Children and Media, 7*(1), 96–113.

Tynes, B. M. (2007). Internet safety gone wild? – Sacrificing the educational and psychosocial benefits of online social environments. *Journal of Adolescent Research, 22*(6), 575–584.

Valcke, M., Bonte, S., De Wever, B., & Rots, I. (2010). Internet parenting styles and the impact on Internet use of primary school children. *Computers & Education, 55*(2), 454–464.

Valcke, M., De Wever, B., Van Keer, H., & Schellens, T. (2011). Long-term study of safe Internet use of young children. *Computers & Education, 57*(1), 1292–1315.

Valkenburg, P. M., & Peter, J. (2007). Preadolescents' and adolescents' online communication and their closeness to friends. *Developmental Psychology, 43*(2), 267–277.

Van den Eijnden, R., Spijkerman, R., Vermulst, A. A., van Rooij, T. J., & Engels, R. (2010). Compulsive internet use among adolescents: Bidirectional parent-child relationships. *Journal of Abnormal Child Psychology, 38*(1), 77–89.

Wang, R., Bianchi, S. M., & Raley, S. B. (2005). Teenagers' Internet use and family rules: A research note. *Journal of Marriage and the Family, 67*(5), 1249–1258.

3
The Role of Culture in Online Behaviour

Alison Attrill

3.1 Culture and online behaviour

When reading the title of this chapter, the question that springs to mind is why think about the role of culture in the applied aspects of online behaviour? In order to answer this question, take a look around you. What do you see? You may be in your own living room, in an office, on a train or a bus. Take a moment to think about these surroundings and other locations in which you live your life. This is the *society* and *culture* within which you currently exist. These surroundings and the people in them influence and are impacted upon by the ways in which you carry out your life, both online and offline. Prior to delving into this further, culture will be defined for the purposes of this chapter.

3.1.1 Defining culture

Culture is "... the collective programming of the mind that distinguishes the members of one group or category of people from others" (Hofstede & Hofstede, 2005, p. 4).

Whilst any given culture is an amalgamation of various norms such as morals, beliefs, and attitudes that shape behaviour, a society is the social grouping to which a person belongs, as defined by its geographical location. It represents the patterns of human behaviour, beliefs, thoughts, opinions, and knowledge passed from generation to generation to shape individuals and societies. You may therefore be of an English society by virtue of living in the geographical location of England, but be a member of a Catholic, Jewish, or Hindu culture because of the cultural groups to which you belong. It is important to note that cultures are formed based on many labels, from religion to sporting preferences. It might be the case that you have always existed within the same society, or that you have transgressed cultures to live in a completely different society from that in which you were born and/or raised. Cultures in which people are raised and those within which they currently exist influence their perceptions and interpretations of who they are. They affect many aspects of behaviour, from the cognitive to the social psychological and many areas in between. In research that focuses on the offline world, these areas are treated as seemingly distinct. The social psychological effects of crowding influence on an office layout might for example be a research focus, or the influence of individual and cognitive factors on choices of education system, job selection, or housing preference. There is a raft of different areas of research that consider the applied aspects of these diverse areas in offline psychology. When considering the relationship of cultural and societal aspects of online behaviour, the distinction between areas of psychology are more blurred. This may partly be due to the rapidly changing environment of the Internet.

3.2 Changing landscapes and gadget overload

The Internet is rapidly changing worldwide in growth and use. Whilst this creates an interconnected world, it makes it extremely difficult to find a consistent flow of theory and research relating to cross-cultural and socio-cultural aspects of online behaviour. Some of the research cited in this chapter may thus appear a little dated, but it seems to currently be the research which can best be used to explore this area of online behaviour. The uptake of Internet use began in Western cultures and has gradually filtered out worldwide (see www.Internetworldstats. com for a range of statistics on worldwide Internet uptake and usage). This might suggest the Internet to have been a Western tool before it even reached some parts of the world. The way in which people in some cultures use the Internet may therefore be overly influenced by Western values and norms, in part due to these having traditionally dominated website design. Another factor that makes drawing conclusions in this area somewhat difficult is that worldwide cyber technologies are not limited to the World Wide Web. Smartphone use has, for instance, rapidly increased over the last five years in many parts of the world, with texting and video-applications such as WhatsApp and Skype rapidly changing the ways in which humans communicate. They are no longer bound by the time delays of asynchronous text communication, but can interact synchronously worldwide. Whilst the relevance of these and other technologies may appear to be ignored in this chapter, the brevity of the work calls for a focus on research and theory that is available to help elucidate cultural factors in Internet use. Applied areas of Internet use such as website design, learning, e-commerce, and shopping will therefore be considered in the remainder of this chapter. Prior to moving to these explorations however, let's ponder the actual term World Wide Web. In order for the Internet to live up to its name of a globally interconnected network of communication and interaction, there needs to be some consistency in the way that it works, regardless of who is using it, where and for what. Nonetheless, some inter-and intra-cultural differences are prevalent in this usage.

3.3 Different cultures

Traditionally, offline psychological research distinguishes between Westernised and non-Westernised cultures. The labels given to these range from individualistic and collectivist, to independent and interdependent cultures, but the underlying premises remain the same: in Westernised cultures such as the UK, Europe, Americas, and Australia, people exist in groups whilst being primarily motivated by their individual needs, achievements, and well-being. In non-Westernised cultures people exist in groups and work for the collective, the group's achievements and well-being. One of the leading theories from which these differences have been derived is Hofstede's (1997) outline of five dimensions of behaviour: *power distance, uncertainty-avoidance, individualism-collectivism, masculinity-femininity*, and *long-term orientation*. Hofstede and Hofstede (2005) further proposed that culture consists of *symbols, social orders, attitudes, goals, practices*, and *values*. Practices can be further sub-divided into *symbols, heroes*, and *rituals*, with *symbols* consisting of *language, words, pictures*, and *gestures*. Whilst there may be cultural heroes (e.g., Mother Theresa), these can also vary within cultures. My heroes might be different to those who you hold in high esteem, for instance, but there remain some heroes who have cultural status. The structure of "culture" according to this layered view sees changes in culture occurring more rapidly in the practices than in the core or cultural values. Values are the core, inflexible attitudes that individuals hold. They lend themselves to testing cultural differences and creating cultural value dimensions such as *individualism-collectivism* offline, but in terms of online behaviour, establishing these differences is somewhat more difficult. People the world over are interacting on a daily basis via the Internet. They are experiencing cultural values and norms from afar that they may never have heard of had the Internet not been invented. This is summed up nicely by Hassan's (2008) contribution to the *Digital Media and Society* Series in which he points out that we now exist in an information society that has *"caused the most significant society shift since the Industrial Revolution"* (p. 23). It is a shift that has created a participatory and self-broadcasting culture, with websites such as YouTube and Instagram extending beyond local communities and infrastructures to a worldwide community. There is an inter-reliant relationship between users posting material and others accessing that material. Of interest for the remainder of this chapter is therefore a) how any given culture shapes and influences such online interactions and b) how these interactions influence and shape society and culture on a wider level. Underlying these interactions is a host of websites which are used to create worldwide links. Attention therefore now turns to considering the role that their design might play in online behaviour.

3.4   Website navigation and design

Website navigation via hypertext or hypermedia to move through a network of information on the Internet is one of the most basic and applied aspects of online behaviour. Regardless of the end result that a user wants from their online activity, there will be some website navigation involved in reaching that goal. Based on Hofstede's (1997) factors that shape cultural attitudes and behaviour (*values, heroes, rituals,* and *symbols*), Luna et al. (2002) put forward a flow model of cross-cultural and cognitive aspects of website navigation. Based on *symbols*, they discuss deGroot's (1992) conceptual feature model as a flow-type model in which language is related to culture-specific conceptual features. Symbols are cultural features recognised the world over. For instance, if you saw a picture of a man in a bowler hat with a briefcase and umbrella, you might instantly recognise this as a symbol of the UK, or a man in lederhosen with a large tankard of beer as a symbol of Germany. From a cognitive psychological perspective, flow is the process that guides an individual through the navigation of a series of linked tasks that require focused attention and perceived control, amongst other characteristics (Csikszentmihalyi, 2000). Luna et al. (2002) call on various different models, such as the theory of planned behaviour (Fishbein & Ajzen, 1975), that link attitudes and behaviour to illustrate the importance of an interaction of site-specific and individual characteristics in peoples' attitudes towards a website. As with the theory of planned behaviour, in which attitudes feed into intentions to carry out a given behaviour prior to the execution of that behaviour, Luna et al. propose that attitudes towards a website are linked to behavioural intentions via the flow of the website design. To illustrate Luna et al.'s notion of symbols and flow: in the UK, if we think of the symbol of sport, for most people the culture-specific feature of *football* or *rugby* might spring to mind. However, in India, the associated culture-specific feature for the symbol of sport might be *cricket*. Think of these as different mental categories, also called *schemas*, that are organised much in the same way as you might order folders on your computer or laptop. We all have what appears to be an infinite number of schemas based on different information about people, objects and items. These are divided into further sub-categories. You might, for instance, hold the general schema of the symbol *animal*. In the UK, one of the most common house pets is the dog. Therefore, you might immediately go to the sub-category of dogs upon hearing that symbol. Each general schema will then be further divided: dogs might be divided into retrievers, labradors, and beagles (schema-specific categories). Language thus provides a mental structure with which cultural symbols are organised into memory. Now think about websites. Whereas early websites were largely text-based, modern website browsing works in a more categorical flow manner: a viewer navigates through a series of links that takes them through different categories of information. One of the really interesting cross-cultural questions that arises from this is whether different cultures use this type of hierarchical structure in language in different ways to construct websites. If so, how does that affect the usability and applicability of those websites? One way to address this question is to consider the cultural congruency of websites.

Culturally congruent websites are those that feature values and symbols consistent with the website visitor's culture. Partial congruency is evident when only some or one of these cultural factors is homeland congruent. Cultural congruency is achieved through content congruity (culture-specific content) and structural congruity (culture-conforming structures). Content congruity might, for example, include product recommendations from local consumers on e-commerce sites, and structural congruity by ensuring that the website is structured in a way appropriate to the target culture. According to Luna et al. (2002), website developers can thus change the wording of a website to accommodate specific cultural factors in order to draw people to the website and to retain their attention as they browse through the website. This line of reasoning makes practical sense when we consider that our world is one of instant gratification. People no longer want to spend ages browsing through websites for information, but want to readily find the material relevant to activated schemas, that are in turn relevant to current processing goals and desires (Mandler, 1982). With some understanding of culturally congruent website design, website developers can easily offer websites that speak to a person's cultural identity, making them more appealing than those which are culturally incongruent.

If perception, interpretation, and interaction with the layout and content of a website are influenced by culturally congruent values, rituals, heroes, and symbols, there could be a further effect on a whole range of applications, from online banking and shopping to having people engage with online therapies. Let's take an extreme example, of terrorists wishing to recruit new members to their cause. They might have sanctions in place that prevent them from creating a locally-based website that is entirely culturally congruent with their values and rituals. They might, however, be aware of like-minded nationals living in another country who could be easily reeled in to their cause if their website is designed with just enough cultural congruency to attract the attention of those individuals, and structural congruity to retain attention, but also sufficient cultural congruency with the website host nation to avoid local suspicion.

One feature of cultural congruency that is somewhat problematic from an applied point of view is that we may erroneously be assuming

a) that people always value cultural congruency
b) that congruency appears to be reliant on salient attitudes or cognitions, and
c) that people are consciously processing the information in front of them.

In relation to the first of these statements, as the Internet opens up the world of international information across borders and locations, it could go either way. These culturally congruent factors could become either weaker or stronger. If the latter applies, cultural congruency could create very strong divides, the likes of which are usually associated with stereotypical thinking and associated negative behaviours. A consideration of some work around online attitudes and stereotyping might also help us explore points b) and c).

3.5   Attitudes

Yuan et al. (2007) suggest that communication flow is often so fast and complex that much of it is processed without conscious engagement and with the use of some form of heuristics (shortcuts) such as attitudes.

Attitudes are cognitive shortcuts that people use to avoid investing lots of processing capacity into judgements and decisions that they make. The reader is directed to any introduction to social psychology textbook for a full overview of the construction of attitudes, attitudinal processing, and the role of attitudes in offline behaviour given that covering these exceeds the confines of this chapter. Focus here will be given to work that has aimed to demonstrate links between attitudinal processing and cultural aspects of online behaviour.

Fulk et al. (1990) proposed the social influence model (SIM) which suggests that people use the knowledge they have of people in their existing social environments to influence the links they create within a social network. Given that humans are habitual creatures who do not want to waste valuable cognitive resources on assessing and evaluating all of the information coming at them from new or unknown sources, they rely on their held views and attitudes as trusty and reliable resources that avoid the expenditure of cognitive energy. Traditionally, it has been thought that core attitudes are retained and difficult to challenge or change, because changes in attitudinal processing can create a psychological tension or imbalance by disrupting social perceptions and norms of "ideal" practices (Gargiulo & Benassi, 1999). One way in which attitudes may influence applied aspects of online behaviour is that attitudinal processing may make people closed-minded to new information, especially if information comes from unfamiliar cross-cultural sources. This could impact on people not being open to working cross-culturally, especially if culturally incongruent information makes them feel uncomfortable. Alternatively, held attitudes may guide behaviour to ignore attitude-inconsistent information to either avoid or reduce psychological tension and maintain psychological equilibrium. This would however result in an attitudinal bias, with only that material that is attitude-congruent receiving attention. This type of closed-mindedness is a feature of the extreme attitudinal processing believed to underlie stereotyping.

3.6 Stereotyping

Most people identify with certain groups or social networks. This is ingroup identification and has been shown to sometimes become sufficiently strong to reduce peoples' willingness to accept information that stems from sources other than a given ingroup. Husted and Michailova (2002) propose that this resistance is due to outgroup contributions disrupting the stability, familiarity, order, and continuity valued and relied upon by an ingroup. When engaging in online behaviour, it could be mistakenly assumed that people are not influenced by these attitudes and ensuing stereotypes, especially if interacting via a mode of communication that eliminates social cues that could convey an outgroup membership. In the Social Identification Deindividuation Model (SIDE), however, Lea and Spears (1991) suggest that it is the absence of social cues which draws attention to any minor outgroup differences or similarities that may be evident in online communications. By focusing on those cues that are available, even the smallest cultural difference could become sufficiently salient to activate stereotypical views. Imagine for example that you are assessing candidates' applications for a course via their online written communications with you. The only feature that you have upon which to judge them, other than their academic credentials, is their written English. In the absence of any social cues, such as facial gestures or speech intonation, you will likely hone in on any distinguishing factors of their written communication in a way that enables you to identify ingroup (native) and outgroup (foreigner) applications. Your underlying stereotypical and attitudinal processes might result in you subsequently judging outgroup communications differently to how you would judge native applicants. Subsequently, any results from the communication and your judgements of that communication might be explained by you overly focusing on that aspect of the written English which identifies and distinguishes the applicant as a foreigner. Of course, this process could also serve to eliminate stereotypical features of communicators and render everyone equal, so that a reduction in stereotypical processing occurs.

Consider your other interactions online. Almost any website that you visit will contain an abundance of information that could cause you to engage in stereotypical and attitudinal processing. According to the SIDE and SIM models you would focus on attitudinal or stereotype-consistent information. Cho and Lee (2008) examined how pre-existing social networks, ingroup/outgroup boundaries, national culture, and outcome expectancies influenced the flow of information cross-culturally. They found that the social context within which information is exchanged constrains the flow of that information between cross-cultural CMC groups. Their participants were more likely to select information offered by their national ingroup than a foreign outgroup. These exchanges are also influenced by individual outcome expectancies of an information-sharing exercise. That is, when entering online exchanges, people will have some idea of what they want to achieve from that exchange. This implies a role of individual cognition in the process of assessing and interpreting information from a non-national source in a way that could enhance both attitudinal and stereotypical thinking. This, in turn, could cause closed-mindedness to novel information from cross-cultural sources via the Internet. Two questions arise from this: 1) How do these considerations have an applied impact on peoples' interactions with and via the Internet? 2) Does the Internet shape peoples' attitudes and stereotypical beliefs, or do people bring to the Internet already cultur-ally defined attitudes and stereotypical views? One way to explore these questions is to consider cross-cultural cognitive styles.

3.7 Cross-cultural cognitive styles

In their cultural cognition theory Faiola and Matei (2006) proposed that individuals bring their culturally bound cognitive styles, which have been influenced and learned through social and cultural impacts, to their online activities. Cognitive styles reflect individual thinking and processing patterns used by people to perceive, interpret, and respond to their environments. There are many facets to these cognitive styles, from analytical to emotional processing, and from deductive reasoning (i.e., deriving a reason by logically evaluating information) to heuristi-cally processing information. The cultural cognition theory captures cultural differences in exactly these types of cognitive tasks (see also Nisbett & Norenzayan, 2002; Nisbet et al., 2001). It proposes that the ways in which people think, feel, and respond to their environments constitute a cognitive style that is based on learned social responses from social interactions and communications with others in a given culture. People therefore bring their cultural biases, or culturally bound patterns of thinking, to their Internet use in a way that likely influences a) how they use the Internet, b) the types of task that they engage in on the Internet, and c) how they let their Internet use influence, and be influenced by their everyday activities.

Think about the different societies that exist within your current resi-dent culture. If this is the culture in which you have always lived, then you may have very strong cultural norms and values that have been shaped over years. If it is a society to which you have migrated, you may more strongly identify with your birth identity than the culture in which you currently live. Sometimes, cultural birth identity becomes even stronger and can more prominently dominate an individual's cognitive style. Now consider how individuals might actively seek out information online in order to reinforce and maintain such culturally bound cognitive styles. It is feasible that people use the Internet in such a cognitive style-congruent manner in order to service and maintain their cultural identity.

It is, however, worth considering that not everyone within a given society or culture will share cognitive styles. There may be individual differences that also play a role in how cultural input influences individual cognitive styles. Think of the USA, for example, where The Amish resist all modern technology. This is a sub-section of a Westernised culture that is often at the forefront of adopting life-changing technological advances. Thus, although cultures may have different shared cognitive styles that are more or less fully developed by adulthood (Luria, 1971, 1976; Nisbett & Norenzayan, 2002; Vygotsky, 1934/1979, 1932/1989), these styles can also change within a culture. The example of the Amish offers us a good point of reflection on how some societies resist the acceptance of modern technology and the changes brought with it. They do not let the Internet shape any aspect of their existence, because they simply do not use it. Therefore, the Internet does not impact upon all applied aspects of a culture or society, nor are entire cultures influenced by shared cognitive styles. Individual choices and lifestyles also impact upon Internet use.

There are lots of factors that differ both across and within cultures that could determine how the Internet features in peoples' daily lives. These range from the psychological, such as attitudes and intentions shaped by language and social norms, to the simple availability and affordability of the Internet. The Internet makes some behaviours easier than a pre-digital era, such as eliminating the necessity to physically go to the bank to pay a bill. Equally, it makes some acts possible that would not have been logistically possible pre-Internet. For example, communicating simultaneously via video conferencing with people on diverse continents of the world. For some, there has been a shift towards lifestyles entirely influenced by information and software that are no longer tools but an aspect of daily life (Lash, 2007b), with many of the daily tasks engaged in online being guided by cognitive processes that are not given much conscious consideration. Consider, for example, when you log on to any specific website of your choice, a website that you access most frequently and for which you no longer have to think about your log on details, but enter them in an almost non-conscious automated manner. This automation may be an artefact of habitual Internet use. Cultures that are not as au fait with the Internet, or those in which it is not as readily available and accessible for many online activities, may not foster such automated Internet use. That said, it may be Internet providers and website designers, amongst others, who create an automation or discrimination of Internet use.

3.8 Automation and the power of unconscious processing

There has been some work which has attempted to demonstrate links between non-conscious, automated communications and cross-cultural differences in Internet behaviour. Some researchers even suggest that non-human processors could be involved in shaping culturally-specific behaviours. Beer (2009) proposed, for example, that computer algorithms are used to differentiate, categorise, and possibly separate website users in an automated manner, but very little research has considered whether this is the case or whether these processes operate outside of users' conscious awareness (e.g., Graham, 2004). Lash (2007a) considered how algorithmic abilities of the Internet could be used to promote hegemony – the power and dominance of one society or social class over others. He suggested that the Internet could be used to create power in online communications based on software that selectively makes information and material more or less available to certain groups of website users. If this proposition appears somewhat far-fetched, think of any website that you might consult for information on any given topic. Now imagine that you live in a culture where holding certain views might not be acceptable. Internet providers could quite easily restrain accessibility to that information from any device within a given geographical location. If software is being used that filters the availability of information or material, it could be perceived to be taking on human-like discriminatory qualities that might influence how the Internet is being used by any given culture, or sub-groups within that culture. This is a scary prospect! Especially if we consider that one of the most fundamental applications of the Internet in almost any culture is information-seeking. If there are discriminations in the information that is made available, it could influence a wide range of information-seeking behaviours, especially those associated with stereotypical thought patterns.

3.9 Information-seeking

Two of the key applied areas associated with automated information provision and information-seeking behaviours are business and e-commerce. Evidence suggests that people use a decision making heuristic online called the country of origin effect (CoO) (Johansson, 1989), which occurs when peoples' existing stereotypical views about the origins of a product influence whether a person will buy that product. Fong and Burton (2008) have demonstrated the CoO for electronic word of mouth (eWOM) recommendations between Chinese and American online shoppers who had never actually met, but who had communicated via online written forums. This is quite an important cultural distinction. If you go online with stereotypical views about the origins of products, then this could have a significant impact upon your online shopping behaviour. Consider, for example, when you last purchased something on eBay. Did you restrict your purchasing choices to items only available within the UK? If so, this could have been the CoO effect in action. The CoO effect also extends to looking for information about products, with, for example, Chinese consumers being more likely to directly search for known sources of information, whereas Americans favour more their own product experiences and knowledge of those products (e.g., Doran, 2002). If, for instance, a person is looking online for hotel recommendations, according to the CoO effect, those recommendations from one's own culture would be given more weighting. Evidence to this effect comes from Fong and Burton (2008) who explored individualistic and collectivist culture mentalities in three online discussion boards relating to digital photography in the USA (eBay, Yahoo, and Google) and three from China (EachNet, Sina, and Netease). They considered all discussions within the two three-month periods of March to May in 2004 and 2005. Across both periods, Chinese posters were significantly more likely to ask for product recommendations based on information provided by their in-group, whereas American posters valued more individualistic personal opinions. This suggests that stereotypical views that underlie effects such as the CoO are rather persistent and enduring, as is often the way with core attitudes. The other side of this coin is that companies may exploit effects such as the CoO to reel in their customers. Returning to the eBay example, a search for a given item might reveal lots of people from different countries selling the same item, with sellers likely providing information targeted at the culture they are selling to rather than at their own culture. Of course, this could also backfire: if an eBay seller from China, for instance, is targeting a UK audience and their item description is full of spelling and grammar errors, purchasers might be put off buying from them.

Evidence for underlying cultural stereotypes guiding online shopping behaviour by influencing the information that people seek comes from Vuylsteke et al. (2010). Using eye-tracker technology, they considered six factors in the use of bulletin board services in China (*frequency of searching online, goal of online search, types of information sought, types of website used, usage patterns on search engines, and contribution of user opinions*) (p. 310). They noted an organised in-group, an almost clan-like structure to the bulletin boards, as well as differences in the way in which Westerners and non-Westerners use search engines. Chinese participants clicked through Internet sites faster and switched between search engine results more than American and Belgian participants. Whereas Westerners based their judgement on a brief search engine description of a site, the Chinese participants suspended judgement until they had actually accessed a website. Vuylsteke et al. (2010) use Hall's (1976) distinction between high- and low-context cultures to explain their findings. Whereas information is often conveyed via situational or contextual factors in high context cultures, these factors play less of a role in low context cultures. Western Europeans have been shown, for instance, to consider objects from a context-detached analytical perspective, whereas East Asians considered the same objects within their situational context (Masuda & Nisbett, 2001). Thus, in Vuylsteke et al.'s (2010) work, Westerners may have focused on individual search terms or words in line with an analytical approach, whilst Chinese users look at information within its wider context. Vuylsteke et al. suggest that Westerners are so used to focusing on the analytical that processing all of the available information on a website might lead to information overload. They therefore use heuristical processing tools such as attitudes to pinpoint salient or relevant information. This can, in turn, lead to prejudiced processing as seen in the CoO effect. It thus appears that website developers need to ensure that a website is presented in such a way as to speak to intended users' cultural cognitive processing tendencies, especially if it is to be used in e-commerce or to foster business relationships. Another way in which website design and information-seeking behaviours may be influenced by cultural differences is the bandwagon effect.

3.10   Bandwagon effect

In his book *Search Engine Society*, Halavais (2008) outlines the key role that search engine algorithms play in making some web pages appear more popular than others. He outlines how this makes the English language extremely powerful in online behaviour, and highlights how algorithms are frequently manipulated in ways that both violate and censor individual use, for example by restricting the sites that appear for certain search terms or enhancing the visibility of material for other search terms. This type of website manipulation is another way in which information can be made more or less accessible to different cultures or societies, as considered by bandwagon effect. Fu and Sim (2011) provide an overview of heuristics employed to navigate the volume of material that Internet users need to go through to find desired content online. One such heuristic is to use indexes of being popular, either via functions such as a *like* count, stars awarded, or written reviews. To this end, they consider how user-created content attracts attention, and the occurrence of the choice-bandwagon effect. This effect emerges when individuals go along with a consensus, a *jumping on the bandwagon* of popular choice (Simon, 1954), and is often associated with psychological studies of conformity. Online, the effect occurs for instance on YouTube, when individuals are drawn to videos because they have been *tagged* and suggest a popularity through collective approvals. Research has shown that Internet users often base online judgements on superficial aspects such as how many hits a video has, user votes, and popularity criteria. Viewers appear initially to disregard content in their credibility and quality judgements of video content (e.g., Metzger et al., 2010). Individuals and groups alike exhibit this tendency to choose content based on its number of views rather than on its quality (Neuman, 1991; Webster & Phalen, 1997). Whilst not offering a direct comparison of cross-cultural cognitive processing factors, this effect can be used to illustrate how different cultures might manipulate individual attention to online content. It suggests that people are not really paying attention to the material but are guided by some sort of subliminal choice processes. Fu and Sim (2011) identify three reasons why an aggregate bandwagon effect in video selection in particular could occur: a) videos with high viewer counts generate interest; b) video awareness is increased via a larger word-of-mouth audience than videos with low viewer ratings (online word of mouth occurs through email dissemination (e.g., Madden, 2007), hyperlinks being embedded in websites, posted on social networking sites, or via other forms of computer-mediated communication), and c) videos that have high viewer counts will appear nearer the top of search engine results than those with low viewer rates (Cho & Roy, 2004; Pan et al., 2007). These could influence entire swathes of a community if they are portraying a message that wheedles its way into individuals' heuristical processing, because this would then guide which content they subsequently seek out on the Internet. Moreover, if information-seeking behaviour is linked to individuals' educational needs and requirements, both automated processing and features such as the bandwagon effect could have strong influences on applied areas of online behaviour, such as online learning.

3.11 Online learning

Online distance-learning technologies bring learners and educators together from far-reaching corners of the globe. There are only a few studies which consider the role of culture in online learning. Devereaux and Johansen (1994) suggest that engaging people with computer-mediated communication for learning may be more difficult in power-distant (e.g., China) than in low-context (e.g., USA) cultures, due to the former having restrictions and control over interpersonal communications. There are also cultural differences in how students expect to be taught. For example, in African and Asian cultures, students expect a more top-down process led by educators who impart knowledge upon their students than do Western cultures (Olaniran & Agnello, 2008).

Cultural differences also emerge in the medium of learning. Whereas individualistic cultures are more familiar with written formats of educational materials, collectivist cultures are more used to oral exchanges of knowledge. Use of the former and associated unstructured learning practices requires an independence and self-reliance that is not the norm of collectivist cultures that rely on obedience and authority (e.g., Patsula, 2002; Usun, 2004). Anderson (2007) suggests that the Internet enables students to *"co-create their own learning"* (p. 1). Therefore, the Internet should give users from diverse cultural backgrounds the freedom and ability to tailor online learning to their cultural demands and learning styles. Olaniran (2009) argues that using technology for education fosters a transactional control relationship whilst empowering the learner to achieve their full potential. That is, there is a mutual influence between educator and learner. He considers the fine balance of enabling learner-control without relinquishing all educator-control, and cites the UK's Open University as an example of how remote learning can be used as "fun" learning that strikes the right balance of educator-relinquished and student-empowered control. In non-Western cultures and developing countries there are, however, a number of socio-cultural factors that impinge on CMC-accessed learning. These include poverty, digital literacy, outdated technologies, and poor Internet infrastructures (see e.g., Abbott, 2001). Olaniran (2009) also considers the state-ownership of the Internet in China, and how a striving learner would need first to acquire a certain level of technological know-how in order to circumvent imposed government regulations. There is little recent work on cross-cultural features of online learning, but this section raises awareness of how important these considerations may be. We assume that the Internet functions equally for all users. It does not. There are differences in educational resources between and even within cultures. Returning to the notion of technological discriminations, if people are trying to self-educate or follow prescribed syllabi online, automated discriminations or restrictions could be extremely influential in the availability of material. This brings us full circle to the ways in which peoples' learned cognitive styles influence both online and offline behaviour. If using the Internet is part of ongoing education, as it is nowadays in many areas of the world, then it is feasible to assume that individual and cultural cognitive styles are being shaped through online behaviour. Thus, the Internet may indeed be shaping how people behave both online and offline in ways that may be influenced by both inter- and intra-cultural facets of online behaviour and interactions.

3.12 Conclusions

When thinking about this chapter, one question that sprang to mind was: what would happen if we suddenly switched off the Internet for a week? What cross-cultural differences and/or similarities might be apparent if we simply switched the clock back to a pre-Internet era? This is, of course, a hypothetical question that will likely never be answered. The Internet has become such an integral part of our daily lives that many use it without even thinking of it as representing an applied aspect of modern life. By considering cultural differences in online behaviour, we can show that not everyone the world over is reliant on the Internet, nor does everyone the world over use it in the same applied manner. There are both intra- and inter-cultural differences in the applied aspects of online behaviour, only a few of which have been touched upon in this chapter. Hopefully, on reading this, the reader is inspired to consider and explore further areas of cross-cultural research in online behaviour.

3.13 References

Abbott, J. (2001). Democracy@internet.asia? The challenges to the emancipatory potential of the net: Lessons from China and Malaysia. *Third World Quarterly, 22*(1), 99–114.

Anderson, T. (2007). Book review – Control and constraint in E-learning: Choosing when to choose. *International Review of Research in Open Distance Learning, 8*(2), 1–3.

Beer, D. (2009). Power through the algorithm? Participatory web cultures and the technological unconscious. *New Media & Society, 11*(6), 985–1002.

Cho, H., & Lee, J.-S. (2008). Collaborative information seeking in intercultural computer-mediated communication groups: Testing the influence of social context using social network analysis. *Communication Research, 35*(4), 548–573.

Cho, J., & Roy, S. (2004). Impact of search engines on page popularity. *Proceedings of the 13th international conference on the world wide web* (pp. 20–29), New York.

Csikszentmihalyi, M. (2000). *Beyond boredom and anxiety: Experiencing flow in work and play*. San Francisco: Jossey-Bass.

deGroot, A. (1992). Bilingual lexical representation: A closer look at conceptual representations. In R. Frost & L. Katz (Eds.), *Orthography, Phonology, Morphology, and Meaning* (pp. 389–412). Amsterdam: Elsevier.

Devereaux, M. O., & Johansen, R. (1994). *Global work: Bridging distance, culture, and time*. San Francisco, CA: Jossey-Bass.

Doran, K. B. (2002). Lessons learned in cross-cultural research of Chinese and North American Consumers. *Journal of Business Research, 10*, 823–829.

Faiola, A., & Matei, S. A. (2006). Cultural cognitive style and web design: Beyond a behavioral inquiry into computer-mediated communication. *Journal of Computer-Mediated Communication, 11*, 375–394.

Fishbein, M., & Ajzen, I. (1975). *Belief, attitude, intention and behaviour: An introduction to theory and research*. Reading, MA: Addison-Wesley.

Fong, J., & Burton, S. (2008). A cross-cultural comparison of electronic word-of-mouth and country-of-origin effects. *Journal of Business Research, 61*, 233–242.

Fu, W. W., & Sim, C. (2011). Aggregate bandwagon effect on online videos' viewership: Value uncertainty, popularity cues, and heuristics. *Journal of the American Society for Information Science and Technology, 62*(12), 2382–2395.

Fulk, J., Schmitz, J., & Steinfield, C. W. (1990). A social influence model of technology use. In J. Fulk & C. Steinfield (Eds.), *Organizations and communication technology* (pp. 117–139). Newbury Park, CA: Sage.

Gargiulo, M., & Benassi, M. (1999). The dark side of social capital. In R. A. J. Leenders & S. M. Gabbay (Eds.), *Corporate social capital and liability* (pp. 298–322). Boston: Kluwer Academic.

Graham, S. (2004). The software-sorted city: Rethinking the 'Digital Divide'. In S. Grahm (Ed.). *The cybercities reader* (pp. 324–332). London: Routledge.

Halavais, A. (2008). *Search engine society*. Digital media and society series. Cambridge: Polity Press.

Hall, E. T. (1976). *Beyond culture*. New York: Random House.

Hassan, R. (2008). *The information society: Cyber dreams and digital nightmares*. Cambridge, UK: Polity Press.

Hofstede, G. (1997). *Cultures and organizations: Software of the mind*. New York: McGraw-Hill.

Hofstede, G., & Hofstede, G. J. (2005). *Cultures and organizations: Software of the mind* (2nd ed.). New York: McGraw-Hill.

Husted, K., & Michailova, S. (2002). Diagnosing and fighting knowledge-sharing hostility. *Organizational Dynamics, 31*(1), 60–73.

Johansson, J. (1989). Determinants and effects of the use of 'made in' labels. *International Marketing Review, 1*, 47–58.

Lash, S. (2007a). Power after hegemony: Cultural studies in mutation. *Theory, Culture & Society, 24*(3), 55–78.

Lash, S. (2007b). New new media ontology. Presentation at *Toward a Social Science of Web 2.0*, National Science Learning Centre, York, UK, 5 September.

Lea, M., & Spears, R. (1991). Computer-mediated communication, deindividuation and group decision-making [Special issue: Computer supported cooperative work and groupware]. *International Journal of Man Machine Studies, 34*, 283–301.

Luna, D., Peracchio, L. A., & deJuan, M. D. (2002). Cross-cultural and cognitive aspects of web site.

Luria A. R. (1971). Towards the problem of the historical nature of psychological processes. *International Journal of Psychology 6*(4), 259–272.

Luria A. R. (1976). *Cognitive development: Its cultural and social foundations.* Cambridge, MA: Harvard University Press.

Madden, M. (2007, July 25). Online video [Report]. Retrieved from the Pew Internet & American Life Project http://www.pewinternet.org/PPF/r/219/report_display.asp

Mandler, G. (1982). The structure of value: Accounting for taste. In M. Clark & S. Fiske (Eds.), *Affect and cognition: The 17th annual Carnegie symposium* (pp. 30–36). Hillsdale, NJ: Lawrence Erlbaum.

Masuda, T., & Nisbett, R. E. (2001). Attending holistically versus analytically: Comparing the context sensitivity of Japanese and Americans. *Journal of Personality and Social Psychology, 81*, 922–934.

Metzger, M. J., Flanagin, A. J., & Medders, R. B. (2010). Social and heuristic approaches to credibility evaluation online. *Journal of Communication, 60*(3), 413–439.

Neuman, W. R. (1991). *The future of the mass audience.* Boston, MA: Cambridge University Press.

Nisbett, R. E., & Norenzayan, A. (2002). Culture and cognition. In H. Pashler & D. L. Medin (Eds.), Stevens' *Handbook of experimental psychology: Vol. 2: Cognition* (3rd ed., pp. 561–597). New York: John Wiley & Sons.

Nisbett, R. E., Peng, K., Choi, I., & Norenzayan, A. (2001). Culture and systems of thought: Holistic vs. analytic cognition. *Psychological Review, 8*, 291–310.

Olaniran, B. A. (2009). Culture, learning styles and Web 2.0. *Interactive Learning Environments, 17*(4), 261–271.

Olaniran, B. A., & Agnello, M. F. (2008). Globalization, educational hegemony, and higher education. *Journal of Multicultural Educational Technology, 2*(2), 68–86.

Pan, B., Hembrooke, H., Joachims, T., Lorigo, L., Gay, G., & Granka, L. (2007). In Google we trust: Users' decisions on rank, position, and relevance. *Journal of Computer-Mediated Communication, 12*(3), Article 3.

Patsula, P. J. (2002). Practical guidelines for selecting media: An international perspective. Usableword Monitor. Retrieved from http://uweb.txstate.edu/*db15/edtc5335/docs/mediaselection_criteria.htm

Simon, H. A. (1954). Bandwagon and underdog effects and the possibility of election predictions. *Public Opinion Quarterly, 18*(3), 245–253.

Usun, S. (2004). Factors affecting the application of information and communication technologies (ICT) in distance education. *Turkish Online Journal of Distance Education, 5*(1). Retrieved from http://tojde.anadolu.edu.tr/tojde13/index.htm

Vuylsteke, A., Wen, Z., Baesens, B., & Poelmans, J. (2010). Consumers' search of information on the internet: How and why China differs from Western Europe. *Journal of Interactive Marketing, 24*, 309–331.

Vygotsky, L. S. (1979). *Mind in society: The development of higher psychological processes.* Cambridge, MA: Harvard University Press. [Translated from the original text published in 1934.]

Vygotsky, L. S. (1989). *Thought and language.* Cambridge, MA: MIT. [Translated from the original text published in 1932.]Webster, J. G., & Phalen, P. F. (1997). *The mass audience: Rediscovering the dominant model.* Mahwah, NJ: Erlbaum.

Yuan, Y. C., Fulk, J., & Monge, P. (2007). Access to information in connective and communal transactive memory systems. *Communication Research, 34,* 131–155.

4

Natives and Immigrants: Closing the Digital Generation Gap

Edward T. Asbury

4.1 Introduction

A YouTube video was posted by the parent of a one-year-old in 2011, which featured the toddler playing with an iPad. The youngster smiles and laughs as she swipes her tiny fingers across the screen and makes the images magically disappear and morph into new ones. The parent then plays a cruel trick on the child. He swaps the tablet with a non-digital magazine. The magazine has just as many colorful images on the cover as the iPad, but the images do not change with a finger swipe. In just a few short moments, the toddler becomes visibly frustrated and upset. The father cleverly entitles the video, "A magazine is an iPad that does not work." Perhaps an extreme example, but the one-minute video clip provides a brilliant illustration of how technology has fundamentally changed the way we see and interact with the world. The technology surrounding us continues to be a lever for the evolution of human behavior. In particular, Prensky (2001) described the resultant dissemination of technology to younger generations as the creation of "digital natives." According to Prensky, the average young college graduate has spent an estimated 5,000 hours of reading in his or her lifetime. The same young adult reportedly has logged over 10,000 hours playing video games. You can double the last number for television-viewing, while computer games, email, Internet, and text messaging are not included. In essence, today's native from a developed country will grow up immersed in the modern conveniences (and inconveniences) that digital "screen-based" technologies provide.

The foundation of this chapter is based on evidence that there are some differences between young and older users of technology. However, generational differences are often a result of goal-directed behavior. Most of us use technology in varying degrees and for a variety of reasons. Although digital platforms continue to merge (we watch TV shows on our computers and surf the web on our TV sets), we will examine several types of screen-based technologies. As a result, we examine fundamental differences and some surprising similarities between digital natives and those of us who have been immigrants forced to learn much of the "language" we were not exposed to as children.

As children migrate into the school-age years, most already have some experience using computers and other screen-based technologies (e.g. tablets, smart phones, video games). In early grade school, children are creating digital presentations, searching for information on the Internet, and submitting school work online. Young and middle-aged adults use some form of screen-based technology every day. Cars now have built-in digital navigation systems and we can attend business meetings away from the office by Skyping from our laptops, tablets, and phones. Researchers examining the behavioral effects of screen-based technology usage often focus on the amount (quantity) of usage as the key predictor for behavioral outcomes. Perhaps the more important questions should revolve around the examination of *what* we are doing and *how* we are doing it. In turn, we will examine both content and contextual issues pertaining to computer-based screen-time for children and adults. While effective boundary management between online and offline behavior is optimal for good mental and physical health, we will conclude with a discussion about why we should be concerned about technology usage across the lifespan.

4.2 Natives versus immigrants

An exploration of the digital divide in relation to Information and Communication Technology (ICT) should focus on both the frequency and type of engagement that are the focus of online behavior. When we examine the frequency of ICT, there is little support for the idea that the young are more active than the old. In fact, there are recent challenges to the entire digital native versus digital immigrant dichotomy. Margaryan et al. (2011) argue that there is no reliable evidence that our youth uses technology more often than adults. Attrill (2015) suggests we can better understand Internet use, in particular, as being based on our goal-directed motivation. The native and immigrant constructs however, do remain useful as an age demographic frame of reference.

If we step back and examine the early research on Internet usage in the 1990s, the literature commonly describes a dichotomy of haves versus have-nots. The digital divide was a line drawn in the sand between the rich and poor. Research appealed to social inequality as education, income, and social status were linked to Internet use. In turn, the research agenda was based on the foundation of a knowledge gap between users and non-users. However, Paul and Stegbauer (2005) argue that a shifting paradigm is warranted, as information does not always equal knowledge. As most of the globe is now connected via the Internet and other digital technologies, the following sections will target specific technologies and motivational factors that may serve as the source of a digital divide when one exists.

4.3   TV in the digital age: killing time

"Today, over 90% of children begin watching TV regularly before the age of 2 years in spite of recommendations to the contrary ... what was once an expensive technology has become so affordable that there are more televisions than toilets in most developed and developing countries" (Christakis, 2009).

Cyberpsychology does not only examine the effects of the Internet on behavior, but rather all technologies. We begin the chapter with a brief discussion of television-viewing habits, as TV is typically the first screen-based technology humans will interface. On the surface, it may appear rather harmless to allow our children to spend countless hours watching age-appropriate network shows and movies. In fact, there are some studies that suggest certain TV programmes and DVDs can actually be beneficial to kids. When Walt Disney's "Baby Einstein" series was launched for example, improvements in a child's language and logic were promised for those viewing the DVDs. By 2006, sales for the DVD series topped $500 million and was endorsed by nearly 30% of parents who felt the lessons were "good for their brains" (Zimmerman et al., 2007). While empirical studies have found no merit to these claims, the successful marketing campaign was primarily based on parental testimony (Christakis, 2009).

Perhaps the most damning links to TV-viewing are the multitude of health concerns replete in the literature. While children are starving in third-world countries, kids from other parts of the globe are suffering from what many consider an epidemic. The World Health Organisation regards childhood obesity as "One of the most serious global public health challenges for the 21st century" (noo.org.uk, 2015). In both the United Kingdom (UK) and the United States (US), about one in five children aged 10–11 are obese (cdc.gov, 2015). Also, there appears to be no clear digital generation divide between the very young and the elderly in terms of TV-viewing rates. In a recent UK survey, older adults were most likely to identify the television as the technological device they could not live without, compared to portable devices for younger cohorts (ofcom.org.uk, 2013).

In some countries, young men are spending a substantial amount of time online or behind TV sets participating in video games. Symptoms of excessive play appear to be prevalent among adult US men with an average age of around 30 (Northrup & Shumway, 2014). Of the over 20 million reported players of online games worldwide, over two thirds were married adults (Hill, 2011). The very nature of role-playing games promotes substantial time investment, as players may experience a state of flow and become less aware of life outside of the screen. As early as 2004, nearly 80% of US gamers reported a need to neglect major facets of their life in order to uphold their position in an online role-playing game (Griffiths et al., 2004). These same adults reported spending an average of 24.7 hours per week playing online and a primary sacrifice they reported was relationships (see Chapter 11 for more details).

4.4 Flipping the switch: on versus off the grid

Ajzen and Fishbein's (1980) theory of reasoned action model has been used to predict human behavior across a variety of domains. The model is based on how our attitudes and intentions motivate us to act. The technology acceptance model (TAM) was later created as an adaption specific to computer usage behavior (Davis, 1986). The latter makes behavioral predictions based on (1) perceived usefulness, (2) perceived complexity, (3) user attitude, and (4) user intention. When digital generational divides are reported, these models serve as a useful framework for understanding specific differences across the lifespan. For example, in general, older adults tend to report weaker attitudes about using technology and own fewer digital devices (Czaja et al., 2006). Adults 60–91 years old also express less interest, more frustration, and more confusion related to technology compared to younger age groups (Purcell et al., 2013). Despite these differences, more recent studies suggest that older adults are simply slower to adopt new technologies (Zickuhr & Madden, 2012).

While natives began a relationship with digital technology in home and school, immigrants were more likely introduced to the computer at work. Beginning in the 1960s, the industrialized work force has experienced exponential growth in the use of computer technologies. As jobs have required more technological competence, adults began to purchase more computers for the home. Since 2004, most households in the industrialized world have at least one computer hooked up to the Internet and many average over two in parts of the US, UK, Western Europe, Japan, Australia, and others (generatorresearch.com, 2015). Stegbauer (2001) describes the consumer trend as the integration of the work and non-work spheres.

As computers have become as common as TV sets in the home, the line between work and home continues to fade. Modern job demands often require employees to be available outside the workplace at a moment's notice. The exponential increase of email communication has turned this platform from a techno-convenience to an overload nuisance. Employees often complain about the inability to keep up with the constant flow of information while dealing with concurrent tasks. Belotti et al. (2005) describe a kind of *techno-stress* becoming more prevalent and pervasive as adults are locked "into the grid." Common reactions from these stressors include anxiety, frustration, and depression. The combination of adults working from home and the lack of leisure time with friends and family is contributing to a recent surge in work-to-home boundary management.

That said, Internet use data tells us that one aspect of a generational digital divide between children and the elderly is virtually non-existent. While children reportedly spend more time online, they are engaged in similar activities as the elderly population (Paul & Stegbauer, 2005). The absence of the work-sphere category for both cohorts lends more time for entertainment and socializing with friends. Children have yet to be exposed to the techno-stressful work environment, so they are likely to view digital technologies as simply fun, convenient, and essential (recent studies indicate the sudden unavailability of computers and smart phones is actually linked to negative emotions in adolescence (see Osit, 2008 for a review). Higher educated elderly in particular, may have simply lost the professional incentive to deal with online information overload.

4.5   Intergenerational divides: parents and children

Early studies on Internet usage suggested adolescents spend more time online than parents (Albero, 2002). In general, parents reported less interest in using technology and associated the Internet with work rather than leisure and entertainment motives of their children. However, intergenerational differences now appear to vary depending on age and geographic location. Hasebrink et al. (2008), for example, reported that parents are more likely to use the Internet than children. Findings over a three-year period indicated parents from most European countries spent proportionally more time online compared to their children. The largest gap appeared between parents and young children. Even so, this gap appears to be closing, as children younger than eight years old are using the Internet more and more frequently (Holloway et al., 2013).

Attitudes about online privacy continue to be disparate (and perhaps a source of contention) between parents and children (see Chapter 2, "Parenting the Online Child"). As more children and teenagers use digital technologies, parents do not often set limits or monitor online behavior. There tends to be a vast difference between how much time a parent believes his or her child spends online and the amount of time reported by children. For example, the Norton Online Family Report (Foresight, 2015) surveyed nearly 10,000 children and adult Internet users across 11 different countries (Australia, Brazil, Canada, China, France, Germany, India, Italy, Sweden, the United Kingdom, and the United States). In general, parents underestimated the amount of time their child spent online. However, there were also cross-cultural mixed findings about online privacy. In the US, for example, about 60% of parents believed they should have total control over a child's online behavior. Italian parents on the other hand, were more likely to allow autonomy over a child's Internet usage (Foresight, 2015). As nearly 50% of European children use the Internet in their own rooms and one third go online via portable devices, it is challenging for a parent to monitor and regulate use (Duerager & Livingstone, 2015). When a parent does not monitor Internet behavior, of particular concern is a parent's ignorance pertaining to their child's negative online incidents. The Norton Online Family Report indicated about 65% of children had been exposed to a negative online incident while only 45% of parents were aware of the experience.

4.6 Going smart and portable

Digital immigrants will recall the portable communication devices used by the crew in the popular Sci-Fi TV series *Star Trek* in the late 1960s. That futuristic prop turned out to be a mere prototype of the modern day portable cellular phone (cell phone). These phones were once considered a high-end communication device for elite business professionals. Now cell phones (smart phones in particular) are in the hands of approximately 1.75 billion users worldwide, and ownership is predicted to reach nearly 50% of people worldwide by 2017 (emarketer.com, 2015). Obviously, phones that are "smart" imply that they do much more than make and receive phone calls. These powerful devices have all but replaced feature phones (call and text-only devices). In terms of device feature usage, smart phones may provide the best example of a generational *digital divide*.

The international consulting firm Deloitte conducted a global mobile consumer survey (GMCS) to learn more about smart phone use by age demographic (deloitte.com/us/mobileconsumer, 2014). The survey included nearly 35,000 respondents, representing over 20 countries and six different age groups. On the surface, the digital divide between young and old for smart phone usage appears to be closing. The largest increase in smart phone sales across age was individuals over age 55. In fact, current estimates by Deloitte indicate a negligible gap between young and old by 2020. However, while older adults are using smart phones at higher rates, GMCS findings indicate a dearth of smart phone *feature usage* compared to younger users. For example, the number of smart phone users who report never downloading an app is a direct linear trend by age. Only 7.5% of users under 34 years of age reported an "app-download-deficiency" (never downloaded an app to their phones). That percentage continues to rise to approximately 30% by age 65 and older. In turn, the closing of the divide is more likely to be a manifestation of fewer feature phones being available on the market. Deloitte's findings suggest that older adults are purchasing more smart phones while using them in a more limited capacity (i.e. more like a feature phone). Text messaging, for example, peaks in young adulthood, while cell phone calls are preferred by digital immigrants (Ling, 2010; Van Volkom et al., 2013).

Deloitte (2014) suggest a number of viable reasons for seniors using smart phones at only a fraction of their potential capacity. First, the screen size of many phones makes it difficult for older people to see, as visual acuity is declining. Yes, phone manufacturers are trending to larger screen options, but the small icons for phone apps are still difficult to read and distinguish from others on the phone. Second, phone plans incur data usage charges that may be confusing for consumers. Retired seniors on a tight budget may be more averse to downloading and accessing smart phone features that incur a financial cost. Similarly, Deloitte found that 25% of smart phone users over age 55 did not know the limits of their data allowance plan. Perhaps older users simply have fewer apps to choose from that cater to their interests. While there are millions of apps available, those directed at senior populations tend to focus to focus on health and wellness (Deloitte, 2014). Older generations may simply prefer to avoid media that is a reminder of poor health and declining function.

Generational differences in smart phone usage also appear to be linked to perceptions about the device itself. Martinez et al. (2012) interviewed women from five different age cohorts about smart phone use. Digital natives more frequently reported using their device for sharing pictures, videos, and music and that they are less likely to switch them off. Many natives have been using their devices as, "an extension of the body itself, they personalise both the hardware and the software of the device, and they see it as something of their own" (Martinez et al., p. 510, 2012). In contrast, digital immigrants tend to view the device as simply a means of communication with friends and family. While the phone is seen as a device for accessing others across the lifespan, "it is the younger users that actively enact it as a tool for controlling the access that others may have to them" (Martinez et al., p. 510, 2012). Teenage girls, for example, reported the device as a tool for independence and control as parents allowed them to go out unsupervised more often. While parents may view the phone as a way to locate and access their kids while away from home, some girls described the practice of ignoring a parent's calls and blaming poor cell phone coverage as a reason for not responding (Martinez et al., 2012).

The trend of younger generations using more smart phone features was also reported in a comprehensive study examining mobile device usage in the United States (Smith, 2015). The most frequent generation gap for usage included social networking and streaming music and video. For example, 91% of smart phone users between 18–29 years old reported using their device for online social networking (e.g. Facebook), compared to only 55% of adults 50 years and older (pewinternet.org, 2015). Online social networking behavior was second only to text messaging amongst the younger adults. A 44% gap and a 43% gap between younger and older generations were reported for streaming video and music respectively. Also, 93% of smart phone owners in the 18–29-year-old group reported using their device to combat boredom (compared to 55% of users over 50 years old) and nearly 50% as a means of avoiding or interacting with people around them (compared to 15% of the older demographic; pewinternet.org, 2015).

4.7 Going social: for better or worse

"The Internet is becoming the town square for the global village of tomorrow" (Gates et al., 1995).

"As Paul and Stegbauer (2005) point out the focus on information access is misleading. They also highlight that too often the debate equates information with knowledge."

Psychologists have been studying the ways people connect with others for well over a century. We know the typical developmental phases of socialization and attachment thanks to the work of Erik Erikson, John Bowlby, Mary Ainsworth, and many others. Students taking an introductory course in developmental or social psychology will be exposed to research on communication that pre-dates digital technologies by decades. Some may say the best predictor of future behavior is past behavior. For both digital natives and immigrants, however, we are only beginning to understand the web of influence (pun intended). In 1995, less than 1% of homes across the globe had access to the Internet (Internetlivestats.com, 2015). To say that Internet usage has increased over the past 20 years would be a gross understatement. For some degree of context, in 2015, the Internet social networking platform Facebook has become the largest digital nation on the globe, with an astounding 1.4 billion daily users (expandedramblings.com).

Once again, generational digital divides appear most evident in terms of specific goal-driven behavior. According to Dutton et al. (2009), the younger demographic is most interested in online activities related to entertainment and content creation. In contrast, adults in the US report a higher frequency of email usage and searching the Web for health-related information (Jones & Fox, 2009). Further data from the US suggests that adults under the age of 50 are significantly more likely to use social networking sites for staying in touch with old friends, compared to adults over 50 (pewinternet.org, 2015).

While the legal age to join most online social network platforms is 14, the presence of children under age 13 on sites such as Facebook is increasing (Altuna et al., 2013). In fact, a survey of Spanish children between the ages of nine and 13 indicated that 40% of them had a social network profile (Garmendia et al., 2011). While there may be a digital divide between parents and children for online social networking, the gap appears to be culturally dependent. For example, Altuna et al. (2013) found the asynchronous social networking divide between parents and children to be much wider in Southern Europe compared to the western US. A similar pattern was reported for synchronous online communication (e.g. chat rooms), where US parents and teens were nearly twice as likely to use those services (see also Chapter 3 of the current text for further outlines of cultural aspects of online behavior).

In the mid-1990s, some researchers were generally optimistic about the potential of the Internet as a catalyst for human connection. Many argued, as the technology became more available to the masses, that improvement in social relationships would be evidenced. Those from this camp believed the Internet would break the geographic boundaries that hinder offline communication. As a result, people would also be able readily to join groups based on common interests rather than mere convenience (Katz & Aspden, 1997). In opposition, other scholars stressed the possibility of an online platform becoming an inferior medium that would link one to others only superficially. In essence, a counterargument contended that monitors and keyboards could not compete with non-screen-based interaction (Stoll, 1995).

Which camp was right? The debate is not an easy one to settle even after more than two decades of research. Once again, we should examine how we are using the technology in order to better understand the effects of technology on behavior. Facebook, for example, has morphed into much more than a platform to communicate with others since it became available to the masses in 2004. We now log on for entertainment, as we play games and watch cute videos of kittens from embedded YouTube clips. After 10 years of research on social networking sites such as Facebook, we are only beginning to understand how these technologies can be used to enhance our lives. For ageing adults in particular, multiple online resources and support groups may provide an effective means for coping with stress. Research from the 1980s consistently found positive relationships between support groups, mental health, and well-being. Many of the findings suggest coping as a byproduct of sharing and learning (Medvene & Krauss, 1989; Potasznik & Nelson, 1984). In essence, support group participants may enjoy a sense of catharsis when allowed to share experiences with others who have similar stories.

While support groups have been around for decades, there are numerous obstacles that may prevent the potential benefits. Limited resources, for example, may exist for those who live in rural environments (Perron, 2002). Furthermore, the demanding nature of the caregiver role may simply prevent attendance at available meetings. Certain disabilities inherently lead to restrictions on time, finances (some support groups charge a fee), and altered sleep and work schedules (Perron, 2002). However, social networking sites are immune to such limitations and provide an attractive alternative for those who are facing stressors related to physical and cognitive decline. Thus, social networking sites may benefit older adults with chronic stress, as well as the middle-aged adults who are serving as caregivers (see also Chapter 1).

While few studies have addressed the therapeutic potential of social networking sites, research based on Internet usage suggests that such communities may actually be more beneficial in terms of emotional support and self-disclosure than their face-to-face counterparts (Salem et al., 1997). Colvin et al. (2004) surveyed caregivers of adult relatives, who used online support groups. The respondents identified several positive aspects of online communication, including anonymity and time flexibility (Colvin et al., 2004). Likewise, Tanis, Das and Fortgens-Sillmann (2011) found the negative links between caregiver strain and well-being to be reduced as a result of support group websites. Participants in a study of adolescent chronic kidney disease online support reported positive outcomes pertaining to the relative ease of connecting with geographically dispersed individuals with shared experiences (Nicholas et al., 2009). Eichhorn (2008) found constructive and useful support being solicited and received in her investigation of online eating disorder support groups. Finally, Asbury and Hall (2013) found highly engaged users of Facebook reporting lower levels of depression and anxiety compared to less engaged users.

There is an increasing awareness about the potential benefits of online social networks for older adults. Online exchange of problems related to ageing with others in similar life circumstances have been linked to a variety of positive outcomes. Nahm et al. (2010) for example, found discussion-board participants on a hip-fracture prevention website had increased knowledge about the issue. Also, participating in online discussion forums, for adults in general, has been found to relieve stress (Wright, 2000). Barak et al. (2008) suggest the use of online social networking enhances an adult's sense of empowerment. Essentially, the benefits of online support groups are related to (1) psychological benefits of writing, (2) emotional expression, (3) enhancing knowledge, (4) increasing cohesiveness between family and friends, and (5) facilitation of decision-making and volition (Barak et al., 2008).

The outcomes of online social networking are not always positive; privacy and security issues for young users are frequently discussed. As older adults continue to engage in online forums more frequently, similar risks present themselves. Identity theft, for example, may occur when older adults reveal too much personal information. While the crime cuts across generational cohorts, the elderly in cognitive decline are prime targets for perpetrators. Furthermore, Leist (2013) identifies the need for increased consideration of online user rights pertaining to the elderly. For example, the deletion of online content or other modifications of a user's online profile are relevant issues for the family of someone who is suffering from cognitive decline. As more elderly people go online, the ageing cohort is at higher risk of exploitation and fraud.

4.8   Conclusion

While there are some differences in technology use across the lifespan, the general notion of a generational digital divide is losing support. There may be the same variation in skill related to technology within cohorts compared to between cohorts (Bennett et al., 2008; Kennedy et al., 2006). When a gap between cohorts does occur, the division is often mediated by culture and individual goal-driven behavior. Much of the early research on the generational digital divide focused on increasing the use of digital technologies in the elderly cohort. However, Wei (2005) reminds us that the elderly are not a homogeneous group. Some are healthier than others, more adventurous, and more open to new practices and products. There are also differences in cognitive age; today's person over the age of 50 often feels younger than their chronological age (Peral-Peral et al., 2015). Moreover, today's elderly have been exposed to digital technologies for longer than previous generations. Now people between 59 and 69 years old typically report lower anxiety about using technology in general than the same cohort only several years ago (Agudo et al., 2012).

So why are we even concerned that a digital divide exists? Working professionals often savor time away from technology and celebrate "going off the grid" for long weekends and summer vacations. Techno-stress appears to be a real phenomenon, contributing to poor mental and physical health. Cyberbullying and Internet addiction have been the topics of multiple focus groups and conferences. The industrialized world communicates more in front of a screen than face-to-face.

Perhaps we want to bridge digital gaps because, as cyberpsychologists, we firmly believe there are many ways we could benefit and enhance our lives by using more technology. We can virtually replicate face-to-face communication long-distance via the miracle of Skype. Online support groups are abundant and immune to many of the challenges of face-to-face meetings. Social networking sites like Facebook allow us to stay in touch with our friends and families via the advantage of asynchronous communication. Are all of these technologies a preferred substitute for non-digital face-to-face communication and interaction? Probably not, as we may prefer non-computer-mediated interaction in times of critical adverse life events (e.g. Lewandowski et al., 2011). However, as we become a more global society, it seems a worthwhile endeavor to explore ways to bridge digital gaps to enhance overall well-being.

4.9   References

Agudo, S., Pascual, M. A., & Fombona, J. (2012). Usos de las herra-mientas digitales entre las personas mayors. *Comunicar, 20,* 193–201.

Ajzen, I. & Fishbein, M. (1980). *Understanding attitudes and predicting social behavior.* Englewood Cliffs, New Jersey: Prentice Hall.

Albero, M. (2002). Teenagers and internet. Myths and realities of the information society. Retrieved March 14, 2015 from http://www.ehu.es/zer/zer13/adolescent3es13

Altuna, J., Aydin, H., Ozfidan, B., & Amenabav, N. (2013). The digital user of social networks: A comparative, transcultural and intergenerational study. *International Online Journal of Educational Sciences, 5*(3), 598–610.

Asbury, T., & Hall, S. (2013). Facebook as a mechanism for social support and mental health wellness. *Psi Chi Journal of Psychological Research, 18*(3), 124–129.

Attrill, A. (2015). *Cyberpsychology.* Oxford, UK: Oxford University Press.

Barak, A., Boniel-Nissim, M., & Suler, J. (2008). Fostering empowerment in online support groups. *Computer & Human Behavior, 24,* 1867–1883.

Bellotti, V., Ducheneaut, N., & Howard, M. (2005). Quality versus quantity: E-mail-centric task management and its relation with overload. *Human-Computer Interaction, 20,* 89–138.

Bennett, S., Maton, K., & Kervin, L. (2008). The digital natives debate: A critical review of the evidence. *British Journal of Educational Technology, 39*(5), 775–786.

Centers for disease control and prevention. (2015). Retrieved March 14, 2015 from http://www.cdc.gov

Christakis, D. A. (2009). The effects of infant media usage: What do we know and what should we learn? *Acta Paediatrica, 98*(1), 8–16.

Colvin, J., Chenoweth, L., Bold, M., & Harding, C. (2004). Caregivers of older adults: Advantages and disadvantages of internet-based social support. *Family Relations, 53*(1), 49–57.

Czaja, S., Charness, N., Fisk, A. D., Hertzog, C., Nair, S. N., Rogers, W. A., & Sharit, J. (2006). Factors predicting the use of technology: Findings from the center for research and education on aging and technology enhancement. *Psychology and Aging, 21*(2), 333–352.

Davis, D. (1986). Perceived usefulness, perceived ease of use, and user acceptance of information technology. *MIS Quarterly, 13*(3), 319–339.

Duerager, A., & Livingstone, S. (2015). How can parents support children's internet safety? Retrieved from www.eukidsonline

Dutton, W., Helsper, E., & Gerber, M. (2009). Oxford Internet survey 2009 report: The Internet in Britian. Oxford: Oxford Internet Institute, University of Oxford. Retrieved March 14, 2015 from http://microsites.oii.ox.ac.uk/oxis/publications

Eichhorn, K. C. (2008). Soliciting and providing social support over the internet: An investigation of online eating disorder support groups. *Journal of Computer-Mediated Communication, 14*(1), 67–78.

Foresight, I. (2015). Norton online family report. Retrieved March 14, 2015 from http://us.norton.com/norton-online-family-report/promo

Garmendia, M., Garitaonandia, C., Martinez, G., & Casado, M. A. (2011). Risks and safety on the internet: Spanish teenagers in the European context. Retrieved from www.eukids.online

Gates, B., Myhrvold, N., & Rinearson, P. (1995). *The road ahead*. New York: Viking.

Global mobile consumer survey: US edition | deloitte US | telecommunications industry. (2014). Retrieved March 14, 2015 from http://www.deloitte.com/us/mobileconsumer

Griffiths, M. D., Davies, M. N., & Chappell, D. (2004). Online computer gaming: A comparison of adolescent and adult gamers. *Online Computer Gaming: A Comparison of Adolescent and Adult Gamers, 27*(1), 87–96.

Hasebrink, U., Livingstone, S., & Haddon, L. (2008). Comparing children's online opportunities and risks across Europe: Cross-national comparisons for EU kids online. Retrieved from www.eukidsonline.net

Hill, S. (2011). MMO subscriber populations. Retrieved March 14, 2015 from http://www.brighthub.com/videogames/mmo/articles/35992.aspx

Holloway, D., Green, L., & Livingstone, S. (2013). Zero to eight. Young children and their internet use. Retrieved from www.eukidsonline.net

Home: Public health England obesity knowledge and intelligence team. (2015). Retrieved March 14, 2015 from http://www.noo.org.uk

Internet live stats – internet usage & social media statistics. (2015). Retrieved March 14, 2015 from http://Internetlivestats.com

Internet users from 68 countries. (2015). Retrieved March 14, 2015 from http://www.generatorresearch.com/tekcarta/

Jones, S., & Fox, S. (2009). Generations online in 2009. Washington, DC: Pew Internet & American Life Project Report. Retrieved March 14, 2015 from http://pewinternet.org/~?media//Files/Reports/2009/PIP_Generations.2009

Katz, J. E., & Aspden, P. (1997). A nation of strangers? *Communications of the ACM, 40*(12), 81–86.

Kennedy, G., Krause, K., Gray, K., Judd, T., Bennett, S. J., Maton, K. A., Dalgarno, B., & Bishop, A. (2006). Questioning the net generation: a collaborative project in Australian higher education. In L. Markauskaite, P. Goodyear, & P. Reimann (Eds.), *Annual Conference of the Australasian Society for Computers in Learning in Tertiary Education* (pp. 413–417). Sydney, Australia: Sydney University Press.

Leist, A. (2013). Social media use of older adults: A mini-review. *Gerontology, 59*, 378–384.

Lewandowski, J., Rosenberg, B. D., Parks, M. J., & Siegel, J. T. (2011). The effect of informal social support: Face-to-face versus computer-mediated communication. *Computers and Human Behavior, 27*(5), 1806–1814.

Ling, R. (2010). Texting as a life phase medium. *Journal of Computer-Mediated Communication, 15*(2), 277–292.

Margaryan, A., LITTLEJOHN, A., & Vojt. G. (2011). Are digital natives a myth or reality? University students' use of digital technologies. *Computers and Education, 56*(2), 429–440.

Market research on digital media, internet marketing. (2015). Retrieved March 14, 2015 from http://www.emarketer.com

Martinez, I., Aguado, J., & Tortajada, I. (2012). Gendered mobile appropriation, identity boundaries and social role coordination. *Feminist Media Studies, 12*(4), 506–516.

Medvene, L. J., & Krauss, D. H. (1989). Causal attributions and parent-child relationships in a self-help group for families of the Mentally ill. *Journal of Applied Social Psychology, 19*(17) 1413–1430.

Nahm, E. S., Barker, B., Resnick, B, Covington, B., Magaziner, J., & Brennan, P. F. (2010). Effects of a social cognitive theory-based hip fracture prevention website for older adults. *Computer Information Nursing, 28*(6), 371–379.

Nicholas, D. B., Picone, G., Vigneux, A., McCormick, K., Mantulak, A., McClure, M., & MacCulloch, R. (2009). The evaluation of an online peer support network for adolescents with chronic kidney disease. *Journal of Technology in Human Services, 27*, 23–33.

Northrup, J., & Shumway, S. (2014). Gamer widow: A phenomenological study of spouses of online video game addicts. *The American Journal of Family Therapy, 42*(4), 269–281.

Osit, M. (2008). Generation text: Raising well-adjusted kids in an age of instant everything. New York, NY, US: AMACOM.

Paul, G., & Stegbauer, C. (2005). Is the digital divide between young and elderly people increasing? Retrieved March 14, 2015 from http://firstmonday.org/ojs/index.php/fm/article/view/1286/1206

Peral-Peral, B., Arenas-Gaitan, J., & Villarego-Ramos, A. (2015). From digital divide to psycho-digital divide: Elders and online social networks. *Media Education Research Journal, 45*(25), 57–64.

Perron, B. (2002). Online support for caregivers of people with a mental illness. *Psychiatric Rehabilitation Journal, 26*(1), 70–77.

Potasznik, H., & Nelson, G. (1984). Stress and social support: The burden experienced by the family of a mentally ill person. *American Journal of Community Psychology, 12*, 589–607.

Prensky, M. (2001). Digital natives, digital ommigrants Part 1. *On the Horizon, 9*(5), 1–6.

Purcell, K., Brenner, J., & Rainie, L. (2013). Search engine use 2012. Pew Research Center's internet and American life project. Retrieved March 14, 2015 from http://www.pewinternet.org/Reports/2012/Search-Engine-Use-2012/Main-findings-Collecting-information.aspx

Salem, D. A., Bogar, G. A., & Reid, C. (1997). Mutual help goes on-line. *The Journal of Community Psychology, 25*(2), 189–207.

Smith, A. (2015). Pew Research Center: Internet, Science & Technology. Retrieved from http://pewinternet.org

Stegbauer, C. (2001). *Boundaries of virtual communities: Structures in Internet-based communication fora.* Wiesbaden: Westdeutscher Verlag.

Stoll, C. (1995). *Silicon snake oil: Second thoughts on the information highway.* New York: Doubleday.

Tanis, M., Das, E., & Fortgens-Sillmann, M. (2011). Finding care for the caregiver? Active participation in online health forums attenuates the negative effect of caregiver strain on wellbeing. *Communications, 36*(1), 51–66.

US smartphone use in 2015. (2015). Retrieved March 14, 2015 from http://www.pewinternet.org/2015/04/01/us-smartphone-use-in-2015/

Van Volkom, M., Stapley, J. C., & Malter, J. (2013). Use and perception of technology: Sex and generational differences in a community sample. *Educational Gerontology, 39*, 729–740.

Wei, S. C. (2005). Consumer's demographic characteristics, cognitive ages, and innovativeness. *Advances in Consumer Research, 32*, 633–640.

Wright, K. (2000). Computer-mediated social support, older adults, and coping. *Journal of Communication, 50*, 100–118.

Zickuhr, K., & Madden, M. J. (2012). Older adults and internet use. Pew research Center's internet and American life project. Retrieved March 14, 2015 from http://pewinternet.org/Reports/2012/Older-adulst-and-internet-use.aspx

Zimmerman, F. J., Christakis, D. A., & Meltzoff, A. N. (2007). Television and DVD/Video viewing in children younger than 2 years. *Pediatric and Adolescent Medicine, 161*(5), 473–479.

5

Technology-Assisted Memory

Tom Mercer

5.1 Introduction: technological memory aids

Memory is extremely important. It helps us to perform our everyday tasks, remember facts, and reminisce about the past. It also provides us with a sense of identity and allows us to think about and plan for the future (Baddeley et al., 2015). Yet memory is far from perfect. Psychological research has consistently shown that people are forgetful and the memories we do have are often inaccurate and distorted (see Baddeley et al., 2015). We might have difficulty remembering factual knowledge (like the name of a famous person) or other information we once knew (like events we have recently experienced; Crawford et al., 2003). This represents a problem with *retrospective* memory. Alternatively, we might forget to perform an intended action in the future, such as watching an anticipated TV programme or taking medication (Crawford et al., 2003). This is a failure of *prospective* memory.

As a result of our imperfect memory, it is no surprise that we use different techniques to improve our chances of remembering. We might, for example, depend on calendars and lists, or rely on other people to remember information for us (Intons-Peterson & Fournier, 1986). By doing this, we can "offload" our memories onto external aids in the environment, and increasingly new and emerging technologies can support remembering in this way. These "technological memory aids" include any electronic tool or device that supplements memory. For instance, search engines offer a wealth of information that is instantly available, so answers to almost any factual question can be rapidly obtained without the need to use our own memories. Similarly, the latest mobile and smartphones are equipped with high quality cameras and video recorders, allowing people to capture events from their own life on an almost continuous basis. Technological memory aids are also capable of supporting prospective memory, helping us to remember future activities, tasks, and deadlines. For example, intelligent personal assistants such as Siri and Cortana can provide reminders, Facebook offers notifications whenever a friend has a birthday, and electronic calendars can organise our working and personal lives.

This chapter will discuss how these technological memory aids can both support and hinder remembering. We will begin by examining the ways in which the Internet and other technologies can essentially memorise information for us. Whilst this can have adverse consequences, technology can also be used to improve memory. We will explore some of the major practical applications of technological memory aids, particularly in terms of helping those with memory impairment. Ultimately, this chapter will show how technology can ease the burden of remembering within a variety of applied settings.

5.2 Is technology changing the way we remember?

Technological memory aids have enormous potential, but concerns have been raised about their negative influence (see Smart, 2013). Scientific investigation into this topic is still in the early stages, but a seminal study carried out by Sparrow et al. (2011) showed that computers do affect memory. In the first experiment, participants had to answer a series of easy or demanding yes/no questions. Regardless of the difficulty of the question, participants were likely to think about computers during the test, suggesting that the concept of knowledge has become closely connected with technology. Furthermore, when the questions were especially hard, participants showed an increased tendency to think about search engines, such as Google. In the next experiment, Sparrow et al. asked their participants to read a series of trivia statements and type these into a computer. Half of the individuals were told that the computer would save their entries, whereas the other half was told that their trivia statements had been erased. This dramatically affected memory for the facts. Specifically, those who thought their entries had been deleted had better recall scores than those whose statements had been saved. This important finding suggests that people are willing to offload memories onto a computer, eliminating the need to actually retain the information themselves.

Sparrow et al. (2011) labelled this phenomenon the "Google effect" and it should be considered within any setting that requires knowledge retention. Some evidence has already shown that medical practitioners rely on electronic resources to answer questions (Edson et al., 2010) and the Google effect could have wider consequences for education. For instance, providing all course materials online, as is common in Higher Education, may actually hinder students' learning. Similarly, Yacci and Rozanski (2012) proposed that search engines like Google reduce the need for students to remember information. When completing a coursework assignment, for example, students could potentially use technology to write the report without the need for any real learning. This would involve locating chunks of information using search engines, paraphrasing the details, and then piecing everything together. By adopting such an approach, students could successfully complete assignments without properly remembering the information or engaging with the sources in any meaningful way. If this is happening, and students become more willing to offload memories onto technology, then teachers need to decide upon a course of action. Yacci and Rozanski argue that the impact of search engines must be carefully considered when designing assessments, with the aim being to promote deep learning. But others have radically different ideas. For instance, the head of a UK exam board made the national news following his controversial suggestion that Google should be allowed in exams (BBC News, 2015). He proposed that future assessments should focus on allowing students to show an understanding of the material, rather than relying on their ability to memorise it.

The debate surrounding the effects of technology and memory within educational settings is likely to continue, but Sparrow et al.'s (2011) research does highlight the impact of technology on the way we remember information. However, technology does not only affect memory for facts – it can also influence what we remember about things experienced in our own lives. Henkel (2014) ran an experiment in which participants underwent a guided tour of an art museum, viewing 30 different objects. They were asked to photograph some of these objects, whereas others were merely observed. The following day, the participants completed a memory test about the objects they had seen and there was evidence of a photo impairment effect. Specifically, photographed objects were more poorly remembered than observed objects, with participants remembering fewer photographed objects overall, as well as less detail about them. Fortunately Henkel's second experiment found that the photo impairment effect was eliminated when participants zoomed in on a specific part of an object whilst taking the photo. Henkel suggested that this may be due to the additional attentional and cognitive processes invoked by the zoom function.

Overall though, both the Google effect and photo impairment effect reveal the potential dangers of relying on technology as a memory aid, but the positive effects cannot be ignored. In a third experiment, Sparrow et al. (2011) asked participants to read a series of trivia statements and type them into a computer. Messages presented by the computer implied that some of the statements had been erased, whereas others had been saved. During the subsequent memory test, the participants were shown all of the statements again (some were identical to those presented earlier, whereas others had been changed slightly). Their task was to decide whether each statement was the same as that presented earlier, and whether it had been saved or erased. Replicating the Google effect, participants were better at remembering information that had been erased, rather than saved. Nonetheless, when asked to indicate whether a statement had been saved or erased, participants were more accurate at remembering what had been saved, in comparison to what had been erased. Sparrow et al.'s final experiment used a similar arrangement, but this time participants were led to believe that all of the facts they had typed into a computer had been saved. After a recall test, participants were given some information about each fact and required to report the folder in which the statement had been saved. Surprisingly, they were better at remembering *where* information had been saved, rather than the content of the statement itself. Sparrow et al. argued that people are able to remember how to access information, even if they cannot directly retrieve that information from memory. This highlights a close relationship between memory processes and technology, with individuals being willing to change their memory strategy if the information is likely to be available later. In cases where certain details will be accessible electronically, the best strategy may be to remember where the information can be accessed.

Storm and Stone (2015) have also shown how saving information via computers can facilitate subsequent memory processes. Their participants memorised lists of words contained within electronic documents. The actual documents were opened from an external hard drive, but participants were sometimes given the chance to save these files for additional study. Storm and Stone aimed to determine how the process of saving information affected memory. Their procedure always involved the study of two lists of words, which participants knew would need to be recalled later. After the first list had been studied, it was either saved onto the computer (rendering it accessible later) or closed (rendering it inaccessible). Participants then studied the words in the second file, followed by recall tests for the contents of both files. Intriguingly, participants remembered significantly more words from the second list when they had been able to save the first one, in comparison to trials in which saving was not allowed. Thus, saving some information actually had a positive effect on remembering other information. According to Storm and Stone, saving information reduces memory costs, as it no longer has to be actively retained. This then prevents the saved information from interfering with subsequent learning.

The research shows that people use technology to remember information for them. This might have negative implications in situations where knowledge retention is particularly important, like exams, or remembering how to perform a specific action, such as using a computer function. At other times, technology has a beneficial effect. The discovery that saving information onto a computer boosts memory for other, related information could be applied to education. Encouraging students to save some information could help the encoding and retention of related information, but the saved data may itself be harder to remember in future, so steps would be needed to counteract this (such as ensuring students return to saved information and re-study it). Overall, people offload memories onto technology in a strategic way. The possible applications of this effect have already been discussed within educational settings, but there are a variety of other situations in which technology can be used to extend and support memory. One of the clearest areas of application relates to individuals suffering from memory impairment, which is discussed in the next section.

5.3   Supporting individuals with memory impairment

Memory disorders can occur for a variety of reasons, including brain injury and the onset of degenerative disorders such as Alzheimer's and Parkinson's disease (Jamieson et al., 2014). Memory problems are very serious – they can affect all aspects of a person's life and make it extremely difficult to function effectively (Svoboda et al., 2012). Thus, in order to help individuals suffering from memory impairment, efforts have been made to develop technology that can boost remembering. Of particular relevance are devices collectively known as Assistive Technology for Cognition, or ATC (Gillespie et al., 2012). ATC was created to support cognition more generally, but it has the potential to reduce or offset a variety of memory problems. For example, ATC that supports prospective memory function can prompt a person to carry out a task they might otherwise forget. Whilst more traditional methods like diaries can serve the same purpose, technological prompting devices have a clear advantage because they can automatically alert an individual to the task that needs to be undertaken (Baldwin & Powell, 2014).

Numerous ATC devices can assist prospective memory over the medium- and long-term, including mobile phones and pagers (Jamieson et al., 2014). Wilson et al. (2001) examined the effectiveness of a particular pager system called NeuroPage in people with memory and planning problems. The pager was used over a seven-week period and reminded the users to perform certain target behaviours linked with everyday life (such as collecting children from school). During the baseline phase of the study, the researchers measured the extent to which the participants performed various behaviours. Individuals assigned to Group A were then given the pager, whereas participants assigned to Group B were placed on a waiting list. After five weeks of using the pager, individuals in Group A were much better at carrying out the target behaviours than those in Group B (the mean success rate was 74.47% for Group A, but just 48.18% for Group B). However, in a subsequent phase of the experiment the conditions were switched, so that those in Group A returned the pager, and those in Group B were given the chance to use it. Five weeks later, the participants in Group B were significantly better at achieving target behaviours than those in Group A. Overall, the pager proved to be a successful memory aid. During the preliminary baseline phase of the study, less than half of the everyday tasks were being performed. After using the pager, the success rate rose to over 76%, and other studies have shown the positive effects of this particular device (Emslie et al., 2007; Wilson et al., 2005).

Mobile and smartphones can also alleviate prospective memory failures in individuals with a memory disorder. Reminders sent directly to a smartphone can improve everyday memory functioning and increase individuals' feelings of independence, confidence and mood (Ferguson et al., 2015). The advantages of these sorts of reminders were clearly shown in the case of JA. He reported severe memory impairment and consequently frequently forgot to carry out specific actions. Baldwin and Powell (2014) therefore used Google Calendar to send reminders to JA via text message. These reminders alerted him to important behaviours, such as attending appointments. Compared to his initial baseline data, the text messages proved to be a highly effective means for reducing JA's forgetting.

Whilst single case studies should be treated with caution, Stapleton et al. (2007) tested the effects of using the reminder function on mobile phones in five individuals with memory impairment. Reminders were extremely valuable for two of the participants, but the remaining three individuals did not benefit. They suffered from more severe memory problems and may have required more in-depth, specific support. Microprompting devices have been designed for such occasions, as they help in the planning of everyday activities that involve a number of discrete stages (Jamieson et al., 2014). For example, O'Neill and Gillespie (2008) described the advantages of a device known as GUIDE, or General User Interface for Disorders of Execution. As its name suggests, this tool guides individuals through the steps needed to complete a complex task. It delivers verbal prompts and questions, and it also accepts simple verbal responses from the user. GUIDE was effectively used to support a man suffering from memory impairment caused by dementia (O'Neill & Gillespie, 2008). The system was employed to help him attach his prosthetic limbs – a task that had previously caused him difficulty. In comparison to attempting the task without the device, GUIDE allowed this individual to perform the task more quickly and with fewer errors. O'Neill et al. (2010) reported similar benefits in a larger study, involving eight patients. Initially, all of these individuals were having trouble performing the varied steps needed to attach a prosthetic limb. Yet by using GUIDE, six of the eight patients showed a significant reduction in the number of errors and omissions made during the task, in comparison to a baseline attempt lacking technological support. Similarly, in a major review of the literature, Jamieson et al. (2014) found that prompting devices have a large, positive effect on memory performance. An added advantage is that these technologies reduce the need for intensive carer support, which allows people with memory loss to become more independent (O'Neill et al., 2010). Conversely, less direct carer involvement may also lead to reduced emotional or social support.

Other forms of ATC can boost retrospective recall; allowing people to better remember events from their own lives (this is known as autobiographical memory). SenseCam – a small, wearable camera that automatically photographs events throughout the day – has proved particularly useful (Hodges et al., 2011). SenseCam takes a photo whenever there are sensory changes (for example, alterations in light) or after a certain period of time has passed. The average user can expect 4,000 photographs per day (Whittaker et al., 2012) and these can easily be uploaded onto a computer. Whilst SenseCam only captures still images, it can improve autobiographical recall in people suffering from memory impairment. This was highlighted in Berry et al.'s (2007) case study, which investigated an individual known as Mrs B. She had suffered damage to the medial temporal lobe (a brain region that is particularly important for memory) following limbic encephalitis, which leads to inflammation of the brain. As a result, she showed clear amnesic symptoms. In an attempt to offset these symptoms, Mrs B was asked to use SenseCam to record events over the course of a day. The following day, her husband asked her to try and recall details of what had occurred, and immediately afterwards they reviewed the SenseCam images. Mrs B would typically look at these images three times. Two days later, she was asked to recall the events for a second time (but in the absence of the specific SenseCam images). Mrs B was also asked to recall the event one, two, and three months later. The data were very encouraging. SenseCam proved to be a significantly more effective memory aid than a traditional diary, and it allowed Mrs B to remember details about events over long periods of time. SenseCam has also been shown to improve autobiographical memory in individuals with other disorders, such as Alzheimer's disease (Woodberry et al., 2015). Whilst the positive impact of SenseCam does seem to contradict Henkel's (2014) photo impairment effect, studies using SenseCam usually permit the user to return to the photos and review them (Henkel's participants were not able to do this). When they can do this, photos can facilitate recall.

In summary, technology can be used to help those suffering from memory impairment, but there is scope for improvement. According to Jamieson et al. (2014), technological aids are often not readily available for those who need them, and the current devices are not without flaws. Consequently, there is scope to further develop systems that overcome some of the disadvantages with existing technologies. As Wilson et al. (2001) have emphasised, the use of ATC itself requires memory, yet many of the people who need this technology have memory problems. In Wilson et al.'s pager study, one individual needed two weeks to associate the auditory alert with the pager, and an additional four weeks to understand the interface. Gillespie et al. (2012) also indicate that many supportive technologies are too novel or challenging to use. The GUIDE system, which relies on a speech-based interface, goes some way to overcoming the difficulties linked with more demanding visual interfaces (O'Neill & Gillespie, 2008). Yet microprompting technologies like GUIDE should in theory support the entire range of daily activities, from preparing food and drink, to dressing and exercising (O'Neill et al., 2010).

Consideration also needs to be given to the logistical challenges that ATC create. For example, prompting technologies can automatically send alerts, but somebody still needs to take responsibility for entering the event information in advance. Additionally, ATC is vulnerable to complications that may render them temporarily unavailable – any device using the Internet would be affected by lapses in connectivity, and portable devices are limited by their battery life, which is a central concern for users (Evald, 2015). Similarly, portable devices need to be charged regularly, and the individual would need to remember to do this.

Yet perhaps the biggest challenge is persuading individuals with memory impairment to use ATC. Svoboda et al. (2012) reported that most people with memory problems do not spontaneously use assistive technologies – especially if they have more pronounced memory difficulties. The individual known as JA, discussed above, initially refused to use any memory aids as he viewed them as embarrassing and stigmatising (Baldwin & Powell, 2014). It was only through using a more discrete tool (Google Calendar) that JA was willing to make use of assistive technology. Other individuals suffering from memory disorder have made similar arguments (Baldwin et al., 2011), and ultimately the needs of specific users must be the primary consideration. Specifically, the ATC that is chosen should consider the individual's sense of identity, as well as whether they feel comfortable using the device (Baldwin et al., 2011).

Gillespie et al. (2012) have made two further observations about ATC that usefully highlights the direction for future application of this technology. Firstly, ATC devices need to be accessible everywhere. To achieve this, Gillespie et al. have proposed that greater attention should be given to smartphones – aside from being easily portable, they have the required level of power and sophistication, and could act as a single platform for supporting a variety of memory-related behaviours. Secondly, Gillespie et al. argue that ATC needs to be more ambitious in scope. Technological memory aids have huge potential and could ultimately support highly complex processes, such as recognising faces, voices, or objects. This has still to be achieved, but the effective application of ATC in the future could go some way towards offsetting the debilitating problems brought about by memory impairment.

5.4   Wider applications of technological memory aids

Aside from tackling the difficulties caused by memory disorder, there are broader applications of technological memory aids for people with an intact memory. SenseCam, for example, has potential value in any situation where an objective recording of events is preferable to relying on unassisted memory. A representative situation was identified by Hodges et al. (2011), who suggested that SenseCam could be useful in recording the amount of physical activity a person has undertaken, or the food and drink they have consumed throughout the course of a day.

SenseCam can offer a more reliable index of this sort of information, so it may be better than traditional methods (like self-report question-naires) that depend on memory. In terms of food and drink intake, O'Loughlin et al. (2013) have already shown that monitoring calorie intake using a combination of SenseCam and a food diary yields more accurate information than using the diary alone. This is important, as many people are not aware of the food-related decisions they make and are instead readily influenced by environmental cues, such as plate size (Wansink, 2010; Wansink & Sobal, 2007). By allowing individuals greater insight into their eating behaviour, SenseCam could help those trying to lose weight. Additionally, SenseCam has value for law enforcement and security services. As Hodges et al. (2011) explain, this technology is able to provide a record of specific situations, which could assist the subse-quent recall of staff during debriefings.

There are possible applications within education settings too. Fleck and Fitzpatrick (2006) showed how SenseCam could be used to facilitate students' reflection on their learning experiences. Their participants were taken on a field trip to a video arcade and used SenseCam to record the event. After the visit, participants reflected on their own learning with a peer. The SenseCam images proved beneficial and acted as a facilitator for discussion. They were also used as an objective record of what happened, which was valuable when the participants' memories disagreed. Similarly, SenseCam can be used to support the reflection of teachers and tutors when discussing a lesson (Fleck & Fitzpatrick, 2009).

These examples relate to very specific situations, but they can all be viewed as part of a broader phenomenon known as lifelogging. This is the process of digitally recording every aspect of daily life in an auto-matic and unobtrusive fashion, essentially creating a complete archive of personal information (Doherty et al., 2012; Whittaker et al., 2012). An example of a lifelogging system is MyLifeBits. This electronic data-base can store almost everything, including documents a person has read, actions they have performed on a computer (such as the websites that have been visited), and phone calls and conversations the user has had (Gemmell et al., 2006). When used in conjunction with SenseCam, MyLifeBits can record many parts of a person's life in digital format (Bell & Gemmell, 2007). Social media can also support lifelogging. Facebook, for instance, allows users to post comments about their thoughts and feelings at a specific moment in time, describe events they have experienced, and communicate with friends (Nadkami & Hofmann, 2012). Similarly, many people seem to enjoy documenting their lives through photos/videos, and there is a strong motivation for sharing these recordings on social media (Eftekhar et al., 2014). The rich data from social media can then be used to encourage reminiscence. Cosley et al. (2012) describe a tool (Pensieve) that was developed for this very purpose. It sends memory triggers (content from social media) to users via email, and these triggers are designed to encourage people to think about and reflect upon the past.

Lifelogging can therefore act as another technological memory aid. By keeping a digital record of a person's life, it would be possible for users to look back on events and remember more information than would otherwise be possible. Furthermore, the reminiscence and reflection promoted by lifelogging can have important psychological benefits. As Doherty et al. (2012) explain, lifelogging supports nostalgia, which can improve mood and self-esteem, and increase feelings of social connectedness. Similarly, the majority of Cosley et al.'s (2012) participants were positive about Pensieve, especially when the memory triggers allowed people to reminisce about experiences they wanted to remember (particularly unique events), or when the trigger was specific enough to boost recall. Lifelogging also has potential beyond helping an individual's memory, as it can provide a legacy that allows people to be remembered after their death (Caprani et al., 2014). The vast amount of information collected through lifelogging could allow "technological heirlooms" to be created, providing a digital record of a person and their memories that can be passed on to the individual's family or friends (Banks et al., 2012).

Nonetheless, there are some challenges that need to be considered. Aside from ethical concerns, the uptake of lifelogging technologies has been limited (Whittaker et al., 2012). One possible reason for this is the huge amount of data collected by many lifelogging devices, with SenseCam being a particularly good example. Kalnikaitė et al. (2010) carried out an experiment in which participants with an intact memory wore SenseCam over a period of two weeks. Their memory for the events was tested two months later, with participants being permitted to use SenseCam images when answering some questions, but not others. Surprisingly, access to SenseCam did not improve the total number of events that were remembered. Part of the difficulty may have been the very large number of photographs available, and the challenge of organising them in a meaningful way (Whittaker et al., 2012). Other research has shown that people often have trouble locating personally relevant information on their computers, such as photos, and this problem is exacerbated by peoples' reluctance to delete digital information (see Whittaker et al., 2010).

As such, there is scope to develop lifelogging technologies that better assist the user in organising their data. But as Whittaker et al. (2012) have stressed, lifelogging technologies need to offer more than a simple factual record of a person's life: they need to properly support reminiscence and reflection on the past. Similarly, lifelogging technologies are better viewed as a way to *support* memory, rather than replace it (Whittaker et al., 2012), and it is important to stress that aiming to remember everything is not desirable – forgetting is a crucial component of a functioning memory system (Baddeley et al., 2015). Specifically, being able to forget means that potentially interfering or irrelevant information can be removed, allowing memory to function more effectively. Lastly, lifelogging technologies do not only need to focus on looking backwards – they may be able to enhance prospective memory, as reminiscing about a positive experience can motivate us to seek out this experience again. For example, Biondoillo and Pillemer (2015) found that participants who remembered a positive exercise experience were more likely to engage in exercise in future, in comparison to those who were not asked to recall any exercise memories. So, despite some of the issues mentioned above, the potential of lifelogging technology is exciting and it is likely to be a valuable memory aid in a variety of different contexts.

5.5 Summary and conclusion

The evidence reviewed throughout this chapter shows that we are willing to offload our memories onto technology. At times this potentially has negative consequences, as shown through the Google effect and the photo impairment effect. But in other cases technology can supplement both retrospective and prospective memory. The value of technological memory aids is particularly clear for those suffering from memory impairment, but there are wider applications too. Devices that allow people to reflect upon their past have uses within a variety of settings, including health, education, and security. They can also be used to promote reminiscence and reflection, allowing us to think meaningfully about our past. These technologies work best when they are supporting, rather than replacing, memory, and there is still much scope for further development. However, as the sophistication of technology continues to increase, there will be increasing applications for technological memory aids. Their potential is enormous, and over the coming years we may begin to see more and more advantages of having technology-assisted memories.

5.6 References

Baddeley, A., Eysenck, M. W., & Anderson, M. C. (2015). *Memory* (2nd ed.). Hove, UK: Psychology Press.

Baldwin, V. N., & Powell, T. (2014). Google Calendar: A single case experimental design study of a man with severe memory problems. *Neuropsychological Rehabilitation.* Advanced online publication.

Baldwin, V. N., Powell, T., & Lorenc, L. (2011). Factors influencing the uptake of memory compensations: A qualitative analysis. *Neuropsychological Rehabilitation, 21,* 484–501.

Banks, R., Kirk, D., & Sellen, A. (2012). A design perspective on three technology heirlooms. *Human-Computer Interaction, 27,* 63–91.

BBC News. (2015, April 30). *Google 'should be allowed in examinations'.* Retrieved April 30, 2015 from http://www.bbc.co.uk/news/education-32531820

Bell, G., & Gemmell, J. (2007). A digital life. *Scientific American, 296,* 58–65.

Berry, E., Kapur, N., Williams, L., Hodges, S., Watson, P., Smyth, G., Srinivasan, J., Smith, R., Wilson, B., & Wood, K. (2007). The use of a wearable camera, SenseCam, as a pictorial diary to improve autobiographical memory in a patient with limbic encephalitis: A preliminary report. *Neuropsychological Rehabilitation, 17,* 582–601.

Biondoillo, M. J., & Pillemer, D. B. (2015). Using memories to motivate future behaviour: An experimental exercise intervention. *Memory, 23,* 390–402.

Caprani, N., Piasek, P., Gurrin, C., O'Connor, N. E., Irving, K., & Smeaton, A. F. (2014). Life-long collections: Motivations and the implications for lifelogging with mobile devices. *International Journal of Mobile Human Computer Interaction, 6,* 15–36.

Cosley, D., Sosik, V. S., Schultz, J., Peesapati, S. T., & Lee, S. (2012). Experience with designing tools for everyday reminiscing. *Human-Computer Interaction, 27,* 175–198.

Crawford, J. R., Smith, G., Maylor, E. A., Della Sala, S., & Logie, R. H. (2003). The prospective and retrospective memory questionnaire (PRMQ): Normative data and latent structure in a large non-clinical sample. *Memory, 11,* 261–275.

Doherty, A. R., Pauly-Takacs, K., Caprani, N., Gurrin, C., Moulin, C. J. A., O'Connor, N. E., & Smeaton, A. F. (2012). Experiences of aiding autobiographical memory using the SenseCam. *Human-Computer Interaction, 27,* 151–174.

Edson, R. S., Beckman, T. J., West, C. P., Aronowitz, P. B., Badgett, R. G., Feldstein, D. A., Henderson, M. C., Kolars, J. C., & McDonald, F. S. (2010). A multi-institutional survey of internal medicine residents' learning habits. *Medical Teacher, 32,* 773–775.

Eftekhar, A., Fullwood, C., & Morris, N. (2014). Capturing personality from Facebook photos and photo-related activities: How much exposure do you need? *Computers in Human Behavior, 37,* 162–170.

Emslie, H., Wilson, B. A., Quirk, K., Evans, J. J., & Watson, P. (2007). Using a paging system in the rehabilitation of encephalitic patients. *Neuropsychological Rehabilitation, 17,* 567–581.

Evald, L. (2015). Prospective memory rehabilitation using smartphones in patients with TBI: What do participants report? *Neuropsychological Rehabilitation, 25,* 283–297.

Ferguson, S., Friedland, D., & Woodberry, E. (2015). Smartphone technology: Gentle reminders of everyday tasks for those with prospective memory difficulties post-brain injury. *Brain Injury, 29*, 583–591.

Fleck, R., & Fitzpatrick, G. (2006). Supporting collaborative reflection with passive image capture. In *Supplementary Proceedings of COOP '06* (pp. 41–48). Carry-le-Rouet, France.

Fleck, R., & Fitzpatrick, G. (2009). Teachers' and tutors' social reflection around SenseCam images. *International Journal of Human-Computer Studies, 67*, 1024–1036.

Gemmell, J., Bell, G., & Lueder, R. (2006). MyLifeBits: A personal database for everything. *Communications of the ACM, 49*, 89–95.

Gillespie, A., Best, C., & O'Neill, B. (2012). Cognitive function and assistive technology for cognition: A systematic review. *Journal of the International Neuropsychological Society, 18*, 1–19.

Henkel, L. A. (2014). Point-and-shoot memories: The influence of taking photos on memory for a museum tour. *Psychological Science, 25*, 396–402.

Hodges, S., Berry, E., & Wood, K. (2011). SenseCam: A wearable camera that stimulates and rehabilitates autobiographical memory. *Memory, 19*, 685–696.

Intons-Peterson, M. J., & Fournier, J. (1986). External and internal memory aids: When and how often do we use them? *Journal of Experimental Psychology: General, 115*, 267–280.

Jamieson, M., Cullen, B., McGee-Lennon, M., Brewster, S., & Evans, J. J. (2014). The efficacy of cognitive prosthetic technology for people with memory impairments: A systematic review and meta-analysis. *Neuropsychological Rehabilitation, 24*, 419–444.

Kalnikaitè V., Sellen A., Whittaker S., & Kirk, D. (2010). Now let me see where I was: Understanding how lifelogs mediate memory. In *CHI 2010: Proceedings of ACM Conference on Human Factors in Computing Systems* (pp. 2045–2054). New York, USA.

Nadkami, A., & Hofmann, S. G. (2012). Why do people use Facebook? *Personality and Individual Differences, 52*, 243–249.

O'Loughlin, G., Cullen, S. J., McGoldrick, A., O'Connor, S., Blain, R., O'Malley, S., & Warrington, G. D. (2013). Using a wearable camera to increase the accuracy of dietary analysis. *American Journal of Preventive Medicine, 44*, 297–301.

O'Neill, B., & Gillespie, A. (2008). Simulating naturalistic instruction: The case for a voice mediated interface for assistive technology for cognition. *Journal of Assistive Technologies, 2*, 22–31.

O'Neill, B., Moran, K., & Gillespie, A. (2010). Scaffolding rehabilitation behaviour using a voice-mediated assistive technology for cognition. *Neuropsychological Rehabilitation, 20*, 509–527.

Smart, P. (2013). *Understanding the cognitive impact of emerging web technologies: A research focus area for embodied, extended and distributed approaches to cognition.* Retrieved March 6, 2015 from http://eprints.soton.ac.uk/350372/

Sparrow, B., Liu, J., & Wegner, D. M. (2011). Google effects on memory: Cognitive consequences of having information at our fingertips. *Science, 333*, 776–778.

Stapleton, S., Adams, M., & Atterton, L. (2007). A mobile phone as a memory aid for individuals with traumatic brain injury: A preliminary investigation. *Brain Injury, 21*, 401–411.

Storm, B. C., & Stone, S. M. (2015). Saving-enhanced memory: The benefits of saving on the learning and remembering of new information. *Psychological Science, 26,* 182–188.

Svoboda, E., Richards, B., Leach, L., & Mertens, V. (2012). PDA and smartphone use by individuals with moderate-to-severe memory impairment: Application of a theory-driven training programme. *Neuropsychological Rehabilitation, 22,* 408–427.

Wansink, B. (2010). From mindless eating to mindlessly eating better. *Physiology & Behavior, 100,* 454–463.

Wansink, B., & Sobal, J. (2007). Mindless eating: The 200 daily food decisions we overlook. *Environment & Behavior, 39,* 106–123.

Whittaker S., Bergman O., & Clough P. (2010). Easy on that trigger dad: A study of long term family photo retrieval. *Personal and Ubiquitous Computing, 14,* 31–43.

Whittaker, S., Kalnikaitė, V., Petrelli, D., Sellen, A., Villar, N., Bergman, O., Clough, P., & Brockmeier, J. (2012). Socio-technical lifelogging: Deriving design principles for a future proof digital past. *Human-Computer Interaction, 27,* 37–62.

Wilson, B. A., Emslie, H. C., Quirk, K., & Evans, J. J. (2001). Reducing everyday memory and planning problems by means of a paging system: A randomised control crossover study. *Journal of Neurology, Neurosurgery and Psychiatry, 70,* 477–482.

Wilson, B. A., Emslie, H., Quirk, K., Evans, J., & Watson, P. (2005). A randomised control trial to evaluate a paging system for people with traumatic brain injury. *Brain Injury, 19,* 891–894.

Woodberry, E., Browne, G., Hodges, S., Watson, P., Kapur, N., & Woodberry, K. (2015). The use of a wearable camera improves autobiographical memory in patients with Alzheimer's disease. *Memory, 23,* 340–349.

Yacci, M., & Rozanski, E. P. (2012). Student information consumption strategies: Implications of the Google effect. *Proceedings of the 2012 iConference* (pp. 248–253). New York, USA: ACM.

6
Internet Addiction: A Clinical Perspective

Daria Kuss

6.1 Introduction

To date, around 40% of the world population is online. Internet usage has grown almost six-fold over the last decade around the globe. In Korea, 96% of Internet users make use of high-speed Internet connections, in comparison to 78% in the UK and 56% in the US (2012, 2013). Since 2000, the US has more than doubled Internet access and use, and mobile Internet use increased extensively in 2011 (The Nielsen Company, 2012a). These statistics evidence that the Internet has become an integral element in today's society. In 2012, children and adolescents in Australia spent an average of 24 hours online per month, compared with 65 hours for individuals aged 18–24 years, and 25–34 year olds spend more than 100 hours per month online (The Nielsen Company, 2012b). Accordingly, young adults are the most active Internet users and spend roughly three hours per day on the Internet (Kuss et al., 2014a).

As Internet use has increased, research on potential negative consequences of excessive Internet use has accumulated. Starting as early as the 1980s, school counsellors were addressed in order to ensure they are aware that excessive video games playing may lead to "addiction" (Soper & Miller, 1983). In order to gain a comprehensive picture of the clinical perspective on Internet addiction, this chapter will first outline the historical context of relevant research. The ultimate aim is to inform the applied area of clinical practice by elucidating the state of research in the field regarding both the classification of Internet addiction and the relevance of this emerging health problem within the clinical context. First, the current state of empirical knowledge of Internet addiction will be presented, which will be followed by introducing the perspectives of international Internet addiction treatment experts and their experiences with treating Internet addiction and its various manifestations in different contexts and countries. It is anticipated that the presentation of empirical studies together with psychotherapeutic knowledge and experience will offer important insights into how research can be translated into actual clinical practice, furthering our understanding of a newly emerging potential mental health concern.

Nearly 20 years ago, the notion of an *Internet Addiction Disorder* was first referred to, and at the time it emerged as a consequence of the perceived pathologising of everyday behaviours (Goldberg, 1996). Based on traditional substance dependence criteria, as put forward by the Diagnostic and Statistical Manual for Mental Disorders (DSM-IV; American Psychiatric Association, 1994), the individual suffering from *Internet Addiction Disorder* had to experience a minimum of three of the following symptoms over a period of 12 months: tolerance, withdrawal, lack of control, relapse, large amounts of time spent online, negative consequences, and continuation of use despite problem awareness (Goldberg, 1996). Subsequently, Young (1996) and Griffiths (1996) emerged as the first researchers inquiring into Internet addiction empirically. Young (1996) used the American Psychiatric Association's substance dependence diagnosis to develop Internet addiction criteria, and presented the case of a female homemaker who gradually increased her activity in chat rooms due to her rising commitment to virtual communities. At the same time, scholars and early adopters have viewed virtual communities in a positive light, as these communities have offered emotional support and a platform for discussion and sharing information (Rheingold, 1993). In Young's study (1996), the homemaker spent substantial amounts of time online and as a consequence neglected her real life responsibilities and eventually developed withdrawal symptoms. This case exemplified for the first time that the stereotypical view of a young male gamer (Griffiths et al., 2003) had to be discarded, and a female user who sought a sense of belonging and comfort online put in his place.

Griffiths (1996) also published early case studies including both males and females, which were then followed by Young (1998), publishing the first exploratory survey of Internet addiction, including 396 self-reported dependent Internet users who endorsed a minimum of five out of eight criteria adapted from a diagnosis of pathological gambling (American Psychiatric Association, 1994), and 100 non-dependent Internet users. On average, the dependent users spent eight times more hours online relative to the non-dependent users, and used chat rooms and MUDs

(Multi-User Dungeons[1]) more often (Young, 1998). The studies published in the last century can be considered to be the springboard for modern empirical research into Internet addiction, which has since greatly proliferated, particularly since the 2000s.

Research into Internet addiction has been fraught with problems (Kuss et al., 2014a), beginning with the various classifications and terminologies used in the scientific community. These spanned from compulsive computer use (Black et al., 1999), Internet dependency (te Wildt, 2011), pathological Internet use (Morahan-Martin & Schumacher, 2000), problematic Internet use (Davis et al., 2002), and virtual addiction (Greenfield, 1999), to Internet addiction disorder (Ko et al., 2005). Despite the abundance of empirical research since the 2000s, classifying Internet addiction remains difficult given that no gold standard of Internet addiction assessment has emerged. Various review papers on Internet addiction have been published since 2005 (Beard, 2005; Byun et al., 2009; Chou et al., 2005; Widyanto & Griffiths, 2006). Some of the recent reviews have highlighted treatment outcome research (King et al., 2011; Liu et al., 2012) and comorbidity (Ko et al., 2012), and others have focused on the biological basis and the psychological factors involved in Internet addiction aetiology (e.g., Billieux & Van der Linden, 2012; Kuss & Griffiths, 2012b). These reviews emphasise that assessment is still an issue in Internet addiction research, making comparisons across studies difficult, particularly in regards to prevalence rates across countries and participant groups.

A recent systematic review (Kuss et al., 2014a) found dissimilar prevalence rates across current epidemiological Internet addiction research, and this may be suggestive of low validity of the Internet addiction construct across research. Adolescent prevalence rates were found to range between 0.8% in Italy (Poli & Agrimi, 2012) and 26.7% of adolescents in Hong Kong (Shek & Yu, 2012). In adults, prevalence rates varied between 1.0% in Norwegian adults (Bakken et al., 2009) and 22.8% of Iranian Internet users (Kheirkhah et al., 2010). Taken together, the reported differences in prevalence rates can partially be attributed to different classification criteria used across studies, more so than differences between age groups (i.e., adolescents and adults). Kuss et al. (2014a) have identified 19 different Internet addiction questionnaires, and these have been found to use very different criteria for classifying Internet addiction, such as the number of problems experienced (e.g., Beutel et al., 2011), or the amount of online hours (Bener et al., 2011; Mythily

---

[1] Multi-User Dungeons were early text-based virtual environments and can be seen as precursors of the contemporary Massively Multiplayer Online Games.

et al., 2008). Rather than using diagnostic cut-offs, which are relevant for cross-comparisons of research results (First, 2005), some research has made use of a dimensional approach, limiting the possibility to inquire into Internet addiction prevalence (Gamez-Guadix et al., 2012; Meerkerk et al., 2009). If Internet addiction is to be treated as a mental disorder, as implied by the "addiction" label and the inclusion of *Internet Gaming Disorder* in the appendix of the diagnostic manual, a cut-off appears necessary so that clinicians are able to separate clients who do not present with the disorder from clients who can be diagnosed with Internet addiction. This is inherently a question of diagnostic sensitivity and specificity (Kraemer, 2007). Accordingly, an addiction classification requires the knowledge and understanding of clients' presenting problems in the medical and therapeutic context, and assessment must fulfil the function of clinical utility.

Internet addiction (or addiction symptoms stemming from excessive Internet use) seems to be prevalent across different ages, as both adolescent and adult groups may suffer from associated problems (Kuss et al., 2014a). Moreover, even if studies use the same assessment instrument for Internet addiction, they may still adopt different cut-off criteria, begging the question of diagnostic validity. Most probably, the severity of Internet addiction experiences in a person scoring 50 on a 100-point scale is lower than for a person scoring 80 on the same scale, and this adds to the inherent ambiguity in contemporary Internet addiction research. Cut-off points are generally used in the clinical context, and are an integral component of diagnosis of mental disorders as evidenced by the relevant diagnostic manuals, the DSM-5 (American Psychiatric Association, 2013) and the International Classification of Diseases (ICD-10) (World Health Organization, 1992). Cut-off points allow clinicians to make a diagnosis and to keep psychiatric and psychological records of their clients. Moreover, they aid medical and psychiatric research (First, 2005), making them an indispensable tool to understand and evaluate Internet addiction symptomatology. Given the persistent diagnostic ambiguity, Internet addiction research can be viewed as being in its infancy.

6.2 Classification

Irrespective of the nosological ambiguities that surround an Internet addiction diagnosis, based on the strong research support as well as clinical evidence of Internet addiction-related problems, the American Psychiatric Association included *Internet Gaming Disorder* in 2013 in the latest edition of the DSM (DSM-5) as a potential mental health problem that requires further empirical and clinical research (Herold et al., 2012). *Internet Gaming Disorder* is presented in the appendix of the diagnostic manual (American Psychiatric Association, 2013), among other conditions that need additional empirical evidence to be included in the manual as mental disorder. It includes nine criteria, namely preoccupation, withdrawal, tolerance, loss of control, continued use irrespective of problem awareness, neglect of alternative recreational activities, escapism and mood modification, deception, and jeopardisation of relationships and job, clearly situating the behaviour within the new diagnostic entity of *Addiction and Related Disorders*. Five or more of the stated symptoms need to be present over the same 12-month period in order for diagnosis.

The DSM-5 criteria are in line with the conceptualisation of a components model of addiction as put forward by Griffiths (2005), which states that both substance-related and behavioural addictions (including Internet addiction) develop via comparable biopsychosocial processes and share six core components, namely the addiction criteria of salience, mood modification, tolerance, withdrawal, relapse, and conflict. Viewed from the perspective of Internet addiction, the first component – salience – refers to a preoccupation with Internet use, and it includes various domains. On the cognitive level, the individual's thinking revolves around Internet use. On the emotional level, the individual craves using the Internet. Finally, on the behavioural level, other previously important behaviours, such as social engagement, are neglected in order to create the time and space to engage in Internet activities. An individual who is addicted to the Internet may constantly think about the next time they are going to go online, crave for logging on to their favourite games and social networks, and sacrifice time with family and friends for the sake of being online. Second, the component mood modification denotes the individual using the Internet in order to feel better and/or to escape and avoid their real lives. Using the Internet improves their mood and allows them to forget the difficulties of their every-day lives (Young, 2004). Third, tolerance indicates that the individual gradually increases the time and energy they expend on using the Internet relative to the time they initially spent using it, given that increased exposure is needed to recreate the initial positive effect (Tsai & Lin, 2003).

Fourth, withdrawal symptoms can be the result of ceasing/decreasing Internet use due to either internal or external motivations, which leads to negative psychological and physiological consequences. Individuals addicted to using the Internet may experience physiological symptoms, such as weakened immunity and physiological dysfunction (Cao et al., 2011). Psychological symptoms comprise depression and anxiety-related manifestations (Yen et al., 2007). Fifth, conflict includes interpersonal and intrapsychic problems that occur because of excessive Internet use. Individuals addicted to using the Internet may jeopardise their interpersonal relationships with family, friends and colleagues, in order to have more time to spend using the Internet (Liu & Kuo, 2007). Intrapsychic conflict occurs when the individual loses control over their behaviours (Treuer et al., 2001), and this may be one of the key experiences fostering the realisation that external help in the form of psychotherapy may be needed. Sixth, relapse indicates the individual is unsuccessful in disengaging from using the Internet (Griffiths, 2005), encumbering abstinence efforts and moderation of addictive Internet use (Murali & George, 2007). Going through a number of abstinence and relapse cycles appears common in individuals trying to quit their addictive behaviours as evidenced in the relevant addiction recovery literature (Prochaska et al., 1992).

The components model of addiction provides a clear, etiologically and diagnostically relevant understanding of Internet addiction, and contributes to explaining the acquisition, development, and maintenance of both substance-related and behavioural addictions. It is in line with Shaffer et al.'s (2004) syndrome model of addiction, which states both substance-related and behavioural addictions develop through comparable distal antecedecents (i.e., various risk factors, including neurobiological factors and contextual factors) which elevate addiction vulnerability. Given the presence of the identified distal antecedents, proximal antecedents precede the development of an addiction, and can include negative events and stress experiences as well as the sustained use of the substance/engagement in the behaviour. The syndrome model of addiction suggests that addictions develop via similar pathways, and share important characteristics, such as symptoms, addiction history, psychology, sociology, and treatment approaches. Moreover, addictions only differ in their actual manifestation (Shaffer et al., 2004). Given the distal and proximal risk factors, some individuals may develop an addiction to a substance, such as alcohol, whereas others may develop an addiction to engaging in a behaviour, such as Internet use. The addiction components model and the syndrome model of addiction suggest that both substance-related and behavioural addictions should be understood and treated in comparable ways.

Empirical evidence for the components model has been provided by a number of studies assessing behavioural addictions, such as exercise (Griffiths et al., 2005), shopping (Clark & Calleja, 2008), gaming (Lemmens et al., 2009), work (Andreassen et al., 2012), and social networking addiction (Andreassen et al., 2012). Recently, Kuss and colleagues (Kuss et al., 2014b) have validated the components model in the context of Internet addiction using a two sample and two measurement approach. The first sample included over 3,000 adolescents in the Netherlands and Internet addiction was measured using the Compulsive Internet Use Scale (CIUS) (Meerkerk et al., 2009). Salience, withdrawal, mood modification, relapse, and conflict loaded highly on the Internet addiction components factor, but Meerkerk et al.'s (2009) original scale did not include a tolerance item, posing questions regarding the extent to which tolerance should be included in a model of Internet addiction. Relevant research in the context of online gaming addiction (Hellman et al., 2013) has suggested that if tolerance was defined as increasing the amount of time spent gaming with the goal to change how one feels, tolerance appears to be an important feature of online gaming addiction. From this evaluation, it appears that future research needs to investigate the extent to which tolerance applies to behavioural addictions, such as Internet addiction, in a similar way as it applies to more traditional substance-related addictions (Kuss et al., 2014b).

In the second sample, over 2,000 UK university students were included (Kuss et al., 2014b) and Internet addiction was assessed using the Assessment for Internet and Computer Gaming Addiction Scale (AICA-S) (Wölfling et al., 2010). In this sample, salience, withdrawal, tolerance, mood modification, relapse, and conflict were acceptable in explaining the Internet addiction components model and contributed significantly to explaining the Internet addiction components factor. In addition to this, this study found that both the AICA-S and the CIUS appeared to be psychometrically sound. Moreover, the results suggested that the Internet addiction components model is a valid and reliable construct, and both clinical and epidemiological studies would benefit from using a short (five- versus six-item) scale for (pre-)diagnostic assessment of Internet addiction symptomatology (Kuss et al., 2014b).

Based on their validation study of the Internet addiction components model (Kuss et al., 2014b), the authors (Kuss et al., 2014c) moved on to extend the original model by developing a nomological network to understand the nature of the Internet addiction components model better, which required to investigate the statistical and deterministic laws that contribute to it (Cronbach & Meehl, 1955). The construct validity of the Internet addiction components model was validated using Internet addiction-relevant data from the aforementioned large and independent samples of Dutch adolescents and UK university students, which were linked to personality trait measures as assessed via the NEO-Five Factor Inventory (Costa & McCrae, 1992) and the Quick Big Five (Vermulst & Gerris, 2005). The results of this study suggested that in the Dutch adolescent sample, low agreeableness, low conscientiousness, and low emotional stability as well as resourcefulness significantly predicted the Internet addiction components factor. Moreover, in the UK student sample, neuroticism, low agreeableness, low conscientiousness, and extraversion predicted the Internet addiction components factor. Overall, this study established a nomological network for the Internet addiction components model, and suggested that low agreeableness and high neuroticism/low emotional stability predicted the Internet addiction components factor in two independent samples, using two measures (Kuss et al., 2014c). From an epidemiological and prevention perspective, the research indicated that certain individuals (possessing the aforementioned personality traits) could be at higher risk of developing Internet addiction. From a clinical and therapeutic perspective, this research suggests that the Internet addiction components model as assessed via the respective questions in the AICA-S and the CIUS can be used for screening purposes of patients seeking help for their Internet-use related problems. In combination with the identified personality traits, individuals at an increased risk for Internet addiction can be identified and supported adequately from a professional point of view (Kuss et al., 2014c).

6.3   Clinical perspectives

The diagnostic ambiguity of Internet addiction and the current lack of a clear and concrete diagnosis as postulated in the diagnostic manuals complicate the provision of specialised treatment for Internet addiction across Europe and the US. Without a definite diagnosis, healthcare funding is often unavailable, impeding treatment provision. Despite the political difficulties surrounding Internet addiction treatment, a number of specialised treatment centres have emerged across Europe, such as the outpatient clinic for behavioural addictions in Mainz, Germany, and the Capio Nightingale hospital in London, UK, and, in the USA, the inpatient centres RESTART Internet Addiction Recovery Program in Seattle and the recently opened Digital Detox and Recovery Centre in Pennsylvania. These growing efforts indicate that the problem of

Internet addiction is being taken more seriously and the development of treatment approaches and facilities is gradually gaining momentum.

In South-East Asian countries, the situation looks somewhat different, as governments and health care providers have accepted Internet and gaming addiction as a threat to general health (Starcevic, 2012), and have consequently established a number of initiatives to help individuals who require professional help to deal with their excessive behaviours. In South Korea, it is estimated that almost a quarter of children suffering from Internet addiction require hospitalisation (Ahn, 2007), whereas in Japan, "fasting camps" have been established that call for complete abstinence from any technology use (Majumdar, 2013). It appears that in countries where Internet and gaming usage are viewed as more acceptable behaviours in society, the likelihood of problematic behaviours is higher (King et al., 2012), providing a possible explanation for high prevalence rates of Internet addiction-related problems in South-East Asian countries.

In a comprehensive study (Kuss & Griffiths, 2015) including 20 Internet addiction treatment experts across six countries in Europe, the US, and Canada, various risk factors were identified that are shared by individuals who seek treatment for their behavioural problems, which testifies to the negative impact their excessive behaviours have on their lives, as well as the lives of their significant others. These risk factors were arranged across individual, situational, and structural characteristics, and are therefore similar to the factors that have been identified to contribute to the acquisition, development, and maintenance of problematic gambling (Griffiths & Wood, 2000). Individual risk factors included age, gender, and the client's profile. Regarding age, research has indicated that the risk of developing Internet addiction can increase with early Internet use (Guan & Subrahmanyam, 2009), and Kuss and Griffiths' research (2015) indicated that parents make use of the Internet as a "babysitter" in some cases, and this contributes to young individuals using the Internet as a tool for emotion regulation. In addition to this, young Internet users may not have developed relevant cognitive and self-reflection skills, and appear to focus on short-term rewards without considering long-term consequences of their potentially addictive Internet use, and therefore may be considered particularly vulnerable to engaging in risky behaviours and to developing addictions (Casey et al., 2005; Galvan et al., 2006).

In terms of gender, Internet addiction treatment experts furthermore stated unanimously that male gender appears as a particular risk factor for developing Internet addiction and consequent treatment-seeking, and the profile of a client seeking help for their addictive behaviours was often considered to be that of a young male gamer, who is shy and socially anxious (Kuss & Griffiths, 2015). This corroborates previous research which has indicated that males are more likely to experience Internet addiction problems (Leung & Lee, 2012; Tsai et al., 2009). The psychotherapists revealed that young men may engage excessively in online games as these games allow them to be successful and excel in this context, and to engage in competitive behaviours. In addition to this, gaming produces a social context which can be established with increased ease online, to the detriment of real life social relationships. Another important finding that emerged concerning gender as a risk factor, as stated by some of the included Internet addiction treatment experts, appeared to be the fact that in girls, Internet addiction may appear as a "quiet addiction," as girls maintain the illusion of normality for longer, and therefore dysfunctional behaviours are not as evident as they are in males, who are more likely to shut themselves off from their real life environments more completely (Kuss & Griffiths, 2015). Representative data from the German population furthermore indicated that female adolescents aged 14–16 years have a higher prevalence (17.2%) of problematic Internet use than males of the same age (13.7%) (Rumpf et al., 2014). Research also suggests that girls tend to favour online applications that allow them to communicate (Andreassen et al., 2012), suggesting that although males appear to seek help for their behaviours more often, females' problems with excessive Internet use require further examination.

Situational risk factors included neglect experienced by clients, being university students, and a trigger that set Internet addiction into motion. In terms of neglect, the psychotherapists suggested that in some cases, their clients experienced a lack of parental guidance, monitoring and attention, which exacerbated problematic behaviours (Kuss & Griffiths, 2015). Parents appear to be particularly important for adolescents, and their positive parenting attitude, and family communication and cohesion have been identified as protective factors for Internet addiction (Park et al., 2008).

Another situational risk factor for Internet addiction identified by psychotherapists was that a part of their client group consisted of university students (Kuss & Griffiths, 2015). Along with age-relevant developmental tasks such as mastery and autonomy, students were found to experience problems in their transition from being cared for and looked after by their parents and families to becoming independent adults who are able to provide for themselves. University life allows them to use their time flexibly and be outside of parental control, and it also creates a new and unfamiliar environment (Kuss et al., 2013), and this may contribute to the development of problematic Internet use.

Lastly, environmental triggers were those which might contribute to an increased situational risk for developing Internet addiction, as stated by Internet addiction treatment experts (Kuss & Griffiths, 2015). Individuals who are exposed to certain stressors in their lives might use the Internet in order to make themselves feel better, and this corresponds with proximal antecedents as discussed earlier (Shaffer et al., 2004). The specific triggers identified by the psychotherapists were situations in which their clients had been mobbed by individuals in their environments, where this encompassed verbal and physical aggression and rejection. Using conditioning principles, the individual learns that using the Internet makes them feel better and therefore they continue engaging in it irrespective of negative consequences and potential addiction, suggesting a self-medication approach (Kuss & Griffiths, 2015).

Finally, structural risk factors were the different types of Internet applications the clients preferred to use (Kuss & Griffiths, 2015). The psychotherapists interviewed stated that most of their male clients had problems with their excessive use of online games, and it appeared that the games most commonly played excessively were Massively Multiplayer Online Role-Playing Games (MMORPGs). Research suggests that MMORPGs have the highest addictive potential relative to other game types, and this appears to be due to their inherent reward structure, social element, and the challenge and sense of achievement they provide for the individual (Kuss & Griffiths, 2012a; Kuss et al., 2012). Moreover, the motivation to socialise was stated to be relevant in the context of Internet addiction, and peer pressure with a potential fear of being socially excluded may contribute to an increased vulnerability to developing Internet addiction. Over time, using online social networks, connecting and maintaining social circles can increase Internet addiction risk (Kuss & Griffiths, 2011). Internet addiction treatment experts do not agree when it comes to the question of whether it is necessary to establish different diagnoses for potential addictions to using particular Internet applications. Some feel that it is important to distinguish specific usages, whereas others are sceptical, as technology evolves rapidly and a distinction of respective behaviours would require including new diagnoses regularly. Other psychotherapists refer to "polymediomania" when denoting their clients' generalised Internet use problems (Kuss & Griffiths, 2015), borrowing from the diagnostic term of "polytoxicomania," denoting multiple substance dependence (American Psychiatric

Association, 2000). Relevant scientific studies indicate that there is a requirement to be specific when it comes to the actual things individuals do online (Morahan-Martin, 2005), and this is mirrored by the APA's decision to include *Internet Gaming Disorder* (rather than Generalized Internet Use Disorder) in the updated diagnostic manual (American Psychiatric Association, 2013). Given the diagnostic ambiguity in clinical settings, future research should specifically address questions of the specificity and sensitivity of an Internet addiction diagnosis, and establish the extent to which problematic usage behaviours are present across different user groups, such as males and females.

6.4   Conclusion

Over the last couple of decades, Internet usage has become an integral element of life for young adults across the globe. With increased Internet usage, clinicians have witnessed a rising demand for treatment of health and psychiatric problems resulting from excessive Internet use. Recent research has mirrored this demand, and the scientific community has advanced their efforts to assess Internet addiction in various samples around the world. Research into Internet addiction has been fraught with problems, including nosological unclarity and diagnostic ambiguity, which have led to distinct prevalence estimates across countries and samples. Many assessment tools have been established using different criteria to classify Internet addiction, eventually leading the American Psychiatric Association to include *Internet Gaming Disorder* in the appendix of the DSM-5 (2013). Research furthermore indicates that Internet addiction can be understood using the Internet addiction components model (Kuss et al., 2014b), and the presence of the respective criteria in individuals suffering from excessive Internet use situates it alongside other substance-related and behavioural addictions. Moreover, from the reported analyses it appears that Internet addiction treatment experts across Europe, the US, and Canada agree that Internet addiction is a condition that requires professional treatment, and a number of risk factors that increase one's chances of developing such an addiction have been identified. Growing efforts to provide specialised treatment for Internet addiction across the world indicate that the problem of Internet addiction is being taken increasingly seriously and the development of treatment approaches and facilities is gradually gaining momentum. Based on the presented literature, it is concluded that Internet addiction research is still in its infancy. Future research is required to corroborate the APA's criteria for Internet Gaming Disorder and to assess the extent to which it is viable to distinguish between different types of problematic Internet usage. Moreover, differences between genders in excessive usage should be assessed in more detail and depth, possibly using qualitative research frameworks.

6.5 References

Ahn, D. H. (2007). *Korean policy on treatment and rehabilitation for adolescents' internet addiction.* Paper presented at the 2007 International Symposium on the Counseling and Treatment of Youth Internet Addiction, Seoul, Korea.

American Psychiatric Association. (1994). *Diagnostic and statistical manual for mental disorders IV.* Washington, D. C.: American Psychiatric Association.

American Psychiatric Association. (2000). *Diagnostic and statistical manual for mental disorders IV, text-revision.* Washington, D. C.: American Psychiatric Association.

American Psychiatric Association. (2013). *Diagnostic and statistical manual of mental Disorders (DSM-5).* Arlington, VA: American Psychiatric Association.

Andreassen, C. S., Torsheim, T., Brunborg, G. S., & Pallesen, S. (2012). Development of a Facebook Addiction Scale. *Psychological Reports, 110*(2), 1–17.

Bakken, I. J., Wenzel, H. G., Gotestam, K. G., Johansson, A., & Oren, A. (2009). Internet addiction among Norwegian adults: A stratified probability sample study. *Scandinavian Journal of Psychology, 50*(2), 121–127.

Beard, K. W. (2005). Internet addiction: A review of current assessment techniques and potential assessment questions. *Cyberpsychology & Behavior, 8*(1), 7–14.

Bener, A., Al-Mahdi, H. S., Ali, A. I., Al-Nufal, M., Vachhani, P. J., & Tewfik, I. (2011). Obesity and low vision as a result of excessive Internet use and television viewing. *International Journal of Food Sciences and Nutrition, 62*(1), 60–62.

Beutel, M. E., Braehler, E., Glaesmer, H., Kuss, D. J., Woelfling, K., & Mueller, K. W. (2011). Regular and problematic leisure-time Internet use in the community: Results from a German population-based survey. *Cyberpsychology, Behavior and Social Networking, 14*(5), 291–296.

Billieux, J., & Van der Linden, M. (2012). Problematic use of the Internet and self-regulation: A review of the initial studies. *The Open Addiction Journal, 5*(Suppl 1: M4), 24–29.

Black, D. W., Belsare, G., & Schlosser, S. (1999). Clinical features, psychiatric comorbidity, and health-related quality of life in persons reporting compulsive computer use behavior. *Journal of Clinical Psychiatry, 60*(12), 839–844.

Byun, S., Ruffini, C., Mills, J. E., Douglas, A. C., Niang, M., Stepchenkova, S., et al. (2009). Internet addiction: Metasynthesis of 1996–2006 quantitative research. *CyberPsychology & Behavior, 12*(2), 203–207.

Cao, H., Sun, Y., Wan, Y., Hao, J., & Tao, F. (2011). Problematic Internet use in Chinese adolescents and its relation to psychosomatic symptoms and life satisfaction. *BMC Public Health, 11.*

Casey, B. J., Tottenham, N., Liston, C., & Durston, S. (2005). Imaging the developing brain: What have we learned about cognitive development? *Trends in Cognitive Sciences, 9*(3), 104–110.

Chou, C., Condron, L., & Belland, J. C. (2005). A review of the research on Internet addiction. *Educational Psychology Review, 17*(4), 363–388.

Clark, M., & Calleja, K. (2008). Shopping addiction: A preliminary investigation among Maltese university students. *Addiction Research & Theory, 16*(6), 633–649.

Costa, P. T., & McCrae, R. R. (1992). *Revised NEO personality inventory (NEO-PI-R) and the NEO Five-Factor inventory (NEO-FFI): Professional manual.* Odessa, FL: Psychological Assessment Resources.

Cronbach, L. J., & Meehl, P. E. (1955). Construct validity in psychological tests. *Psychological Bulletin, 52*(4), 281–302.

Davis, R. A., Flett, G. L., & Besser, A. (2002). Validation of a new scale for measuring problematic Internet use: Implications for pre-employment screening. *Cyberpsychology & Behavior, 5*(4), 331–345.

First, M. B. (2005). Clinical utility: A prerequisite for the adoption of a dimensional approach in DSM. *Journal of Abnormal Psychology, 114*(4), 560–564.

Galvan, A., Hare, T. A., Parra, C. E., Penn, J., Voss, H., Glover, G., et al. (2006). Earlier development of the accumbens relative to orbitofrontal cortex might underlie risk-taking behavior in adolescents. *Journal of Neuroscience, 26*(25), 6885–6892.

Goldberg, I. (1996). Internet Addictive Disorder (IAD) diagnostic criteria. Retrieved April 17, 2013 from http://www.webcitation.org/query?url=http%3A%2F%2Fwww.psycom.net%2Fiadcriteria.html&date=2013-02-06

Gamez-Guadix, M., Villa-George, F. I., & Calvete, E. (2012). Measurement and analysis of the cognitive-behavioral model of generalized problematic Internet use among Mexican adolescents. *Journal of Adolescence, 35*(6), 1581–1591.

Greenfield, D. N. (1999). *Virtual addiction.* Oakland, CA: New Harbinger.

Griffiths, M., & Wood, R. T. A. (2000). Risk factors in adolescence: The case of gambling, videogame playing, and the Internet. *Journal of Gambling Studies, 16*(2/3), 199–225.

Griffiths, M. D. (1996). Internet 'addiction': An issue for clinical psychology? *Clinical Psychology Forum, 97*, 32–36.

Griffiths, M. D. (2005). A 'components' model of addiction within a biopsychosocial framework. *Journal of Substance Use, 10*(4), 191–197.

Griffiths, M. D., Szabo, A., & Terry, A. (2005). The exercise addiction inventory: A quick and easy screening tool for health practitioners. *British Journal of Sports Medicine, 39*(6), e30.

Griffiths, M. D., Davies, M. N. O., & Chappell, D. (2003). Breaking the stereotype: The case of online gaming. *Cyberpsychology & Behavior, 6*(1), 81–91.

Guan, S.-S. A., & Subrahmanyam, K. (2009). Youth Internet use: Risks and opportunities. *Current Opinion in Psychiatry, 22*(4), 351–356.

Hellman, M., Schoenmakers, T. M., Nordstrom, B. R., & van Holst, R. J. (2013). Is there such a thing as online video game addiction? A cross-disciplinary review. *Addiction Research & Theory, 21*(2), 102–112.

Herold, E., Connors, E., & Moore, T. (2012). *American Psychiatric Association board of trustees approves DSM-V. News release.* Arlington, VA: American Psychiatric Association.

International Telecommunication Union. (2012). Internet users. Retrieved April 24, 2015 from http://www.itu.int/ITU-D/ict/statistics/index.html

International Telecommunication Union. (2013). *The world in 2013. ICT facts and figures.* Geneva, Switzerland: International Telecommunication Union.

Kheirkhah, F., Juibary, A. G., & Gouran, A. (2010). Internet addiction, prevalence and epidemiological features in Mazandaran Province, Northern Iran. *Iranian Red Crescent Medical Journal, 12*(2), 133–137.

King, D. L., Delfabbro, P. H., & Griffiths, M. D. (2012). Clinical interventions for technology-based problems: Excessive Internet and video game use. *Journal of Cognitive Psychotherapy: An International Quarterly, 26*(1), 43–56.

King, D. L., Delfabbro, P. H., Griffiths, M. D., & Gradisar, M. (2011). Assessing clinical trials of Internet addiction treatment: A systematic review and CONSORT evaluation. *Clinical Psychology Review, 31*(7), 1110–1116.

Ko, C. H., Yen, J. Y., Chen, C. C., Chen, S. H., & Yen, C. F. (2005). Proposed diagnostic criteria of Internet addiction for adolescents. *Journal of Nervous and Mental Disease, 193*(11), 728–733.

Ko, C. H., Yen, J. Y., Yen, C. F., Chen, C. S., & Chen, C. C. (2012). The association between Internet addiction and psychiatric disorder: A review of the literature. *European Psychiatry, 27*(1), 1–8.

Kraemer, H. C. (2007). DSM categories and dimensions in clinical and research contexts. *International Journal of Methods in Psychiatric Research, 16*, S8–S15.

Kuss, D. J., & Griffiths, M. D. (2011). Online social networking and addiction – A review of the psychological literature. *International Journal of Environmental Research and Public Health, 8*(9), 3528–3552.

Kuss, D. J., & Griffiths, M. D. (2012a). Internet gaming addiction: A systematic review of empirical research. *International Journal of Mental Health and Addiction, 10*(2), 278–296.

Kuss, D. J., & Griffiths, M. D. (2012b). Internet and gaming addiction: A systematic literature review of neuroimaging studies. *Brain Sciences, 2*(3), 347–374.

Kuss, D. J., & Griffiths, M. D. (2015). *Internet addiction in psychotherapy.* London: Palgrave.

Kuss, D. J., Griffiths, M. D., & Binder, J. F. (2013). Internet addiction in students: Prevalence and risk factors. *Computers in Human Behavior, 29*(3), 959–966.

Kuss, D. J., Griffiths, M. D., Karila, L., & Billieux, J. (2014a). Internet addiction: A systematic review of epidemiological research for the last decade. *Current Pharmaceutical Design, 20*(25), 4026–4052.

Kuss, D. J., Louws, J., & Wiers, R. W. W. (2012). Online gaming addiction? Motives predict addictive play behavior in Massively Multiplayer Online Role-Playing Games *Cyberpsychology, Behavior & Social Networking, 15*(9), 480–485.

Kuss, D. J., Shorter, G. W., van Rooij, A. J., Griffiths, M. D., & Schoenmakers, T. (2014b). Assessing Internet addiction using the parsimonious Internet addiction components model – A preliminary study. *International Journal of Mental Health and Addiction, 12*(3), 351–366.

Kuss, D. J., Shorter, G. W., van Rooij, A. J., van de Mheen, D., & Griffiths, M. D. (2014c). The Internet addiction components model and personality: Establishing construct validity via a nomological network. *Computers in Human Behavior, 39*, 312–321.

Lemmens, J. S., Valkenburg, P. M., & Peter, J. (2009). Development and validation of a game addiction scale for adolescents. *Media Psychology, 12*(1), 77–95.

Leung, L., & Lee, P. S. N. (2012). The influences of information literacy, internet addiction and parenting styles on internet risks. *New Media & Society, 14*(1), 117–136.

Liu, C.-Y., & Kuo, F.-Y. (2007). A study of Internet addiction through the lens of the interpersonal theory. *Cyberpsychology & Behavior, 10*(6), 799–804.

Liu, C., Liao, M., & Smith, D. C. (2012). An empirical review of Internet addiction outcome studies in China. *Research on Social Work Practice, 22*(3), 282–292.

Majumdar, A. (2013). Japan plans 'fasting camps' for Web-addicted children. Tech 2. Retrieved September 2, 2013 from http://tech2.in.com/news/general/japan-plans-fasting-camps-for-webaddicted-children/912284

Meerkerk, G. J., Van Den Eijnden, R. J., Vermulst, A. A., & Garretsen, H. F. L. (2009). The Compulsive Internet Use Scale (CIUS): Some psychometric properties. *CyberPsychology & Behavior, 12*(1), 1–6.

Morahan-Martin, J. (2005). Internet abuse. Addiction? Disorder? Symptom? Alternative explanations? *Social Science Computer Review, 23*(1), 39–48.

Morahan-Martin, J., & Schumacher, P. (2000). Incidence and correlates of pathological Internet use among college students. *Computers in Human Behavior, 16*(1), 13–29.

Murali, V., & George, S. (2007). Lost online: An overview of internet addiction. *Advances in Psychiatric Treatment, 13*(1), 24–30.

Mythily, S., Qiu, S., & Winslow, M. (2008). Prevalence and correlates of excessive internet use among youth in Singapore. *Annals Academy of Medicine Singapore, 37*(1), 9–14.

Park, S. K., Kim, J. Y., & Cho, C. B. (2008). Prevalence of Internet addiction and correlates with family factors among South Korean adolescents. *Adolescence, 43*(172), 895–909.

Poli, R., & Agrimi, E. (2012). Internet addiction disorder: Prevalence in an Italian student population. *Nordic Journal of Psychiatry, 66*(1), 55–59.

Prochaska, J. O., DiClemente, C. C., & Norcross, J. C. (1992). In search of how people change. Applications to addictive behaviours. *American Psychologist, 47*(9), 1102–1114.

Rheingold, H. (1993). *The virtual community: Homesteading on the electronic frontier.* Cambridge, MA: MIT.

Rumpf, H. J., Vermulst, A. A., Bischof, A., Kastirke, N., Gürtler, D., Bischof, G., et al. (2014). Occurence of Internet addiction in a general population sample: A latent class analysis. *European Addiction Research, 20*(4), 159–166.

Shaffer, H. J., LaPlante, D. A., LaBrie, R. A., Kidman, R. C., Donato, A. N., & Stanton, M. V. (2004). Toward a syndrome model of addiction: Multiple expressions, common etiology. *Harvard Review of Psychiatry, 12*(6), 367–374.

Shek, D. T. L., & Yu, L. (2012). Internet addiction phenomenon in early adolescents in Hong Kong. *The Scientific World Journal, 2012*, 104304–104304.

Soper, W. B., & Miller, M. J. (1983). Junk-time junkies: An emerging addiction among students. *School Counselor, 31*(1), 40–43.

Starcevic, V. (2012). Is Internet addiction a useful concept? *Australian and New Zealand Journal of Psychiatry, 47*(1), 16–19.

te Wildt, B. T. (2011). Internet dependency: Symptoms, diagnosis and therapy. In K. Cornelius & D. Hermann (Eds.), *Virtual worlds and criminality* (pp. 61–78). Berlin: Springer.

The Nielsen Company. (2012a). State of the media: US digital consumer report: The Nielsen Company. Retrieved November 4, 2015, from http://www.iab.net/media/file/Nielsen_Digital_Consumer_Report_FINAL.pdf

The Nielsen Company. (2012b). Australian online landscape review: The Nielsen Company. Retrieved November 4, 2015 from https://www.iabaustralia.com.au/uploads/uploads/2015-03/1426604400_e7690d6e91430362be5a3011e95e358e.pdf

Treuer, T., Fabian, Z., & Furedi, J. (2001). Internet addiction associated with features of impulse control disorder: Is it a real psychiatric disorder? *Journal of Affective Disorders, 66*(2-3), 283-283.

Tsai, H. F., Cheng, S. H., Yeh, T. L., Shih, C. C., Chen, K. C., & Yang, Y. C. (2009). The risk factors of Internet addiction – A survey of university freshmen. *Psychiatry Research, 167*(3), 294-299.

Tsai, C. C., & Lin, S. S. J. (2003). Internet addiction of adolescents in Taiwan: An interview study. *Cyberpsychology & Behavior, 6*(6), 649-652.

Vermulst, A. A., & Gerris, J. R. M. (2005). *QBF: Quick Big Five Persoonlijkheidstest Handleiding [Quick Big Five Personality Test Manual]*. Leeuwarden, NL: LDC Publications.

Widyanto, L., & Griffiths, M. (2006). 'Internet addiction': A critical review. *International Journal of Mental Health and Addiction, 4*(1), 31-51.

Wölfling, K., Müller, K., & Beutel, M. (2010). Diagnostic measures: Scale for the Assessment of Internet and Computer Game Addiction (AICA-S). In D. Mücken, A. Teske, F. Rehbein & B. te Wildt (Eds.), *Prevention, diagnostics, and therapy of computer game addiction* (pp. 212-215). Lengerich: Pabst Science.

World Health Organization. (1992). *ICD 10: The ICD-10 classification of mental and behavioral disorders: Clinical descriptions and diagnostic guidelines*. Geneva, Switzerland: World Health Organization.

Yen, J. Y., Ko, C. H., Yen, C. F., Wu, H. Y., & Yang, M. J. (2007). The comorbid psychiatric symptoms of Internet addiction: Attention deficit and hyperactivity disorder (ADHD), depression, social phobia, and hostility. *Journal of Adolescent Health, 41*(1), 93-98.

Young, K. (2004). Internet addiction – A new clinical phenomenon and its consequences. *American Behavioral Scientist, 48*(4), 402-415.

Young, K. (1998). Internet addiction: The emergence of a new clinical disorder. *Cyberpsychology & Behavior, 1*(3), 237-244.

Young, K. S. (1996). Psychology of computer use: XL. Addictive use of the Internet: A case that breaks the stereotype. *Psychological Reports, 79*(3), 899-902.

7

Online Support Groups: Enhancing the User Experience with Cyber-Psychological Theory

Chris Fullwood

7.1 Introduction

> No man is an island,
> Entire of itself,
> Every man is a piece of the continent,
> A part of the main. (John Donne, 1624)

Most of us will be familiar with the well-known expression "a problem shared is a problem halved." Conventional wisdom at least then would seem to suggest that there are distinct psychological benefits associated with unburdening your problems onto others. Perhaps there is something cathartic about getting a few things off your chest. Perhaps discussing your problems with others allows you to gain new insights from someone else's perspective or maybe there is just some comfort in simply knowing that others are there for you when you need them. Whatever the case, no matter how resilient you might think you are, there comes a time in all of our lives when (to quote a very famous song) "we all need somebody to lean on" (Withers, 1972).

If I asked you to dredge your memory banks for either an example of a major incident in your life where you needed support from others or an instance where you were called upon to assist someone else in their time of need, I would bet that you would be hard pushed not to come up with at least one example. Most of you will likely conjure up abundant cases. This would almost certainly be true if you (or a close friend or family member) had experienced one or more of the major life events listed on Holmes and Rahe's (1967) Social Readjustment Rating Scale. Divorce, marital separation, the death of a close family member, marriage, pregnancy, and dismissal from work represent just a few examples of major life events where support from others in some shape or form is likely to be required to help you through. Research also suggests that there is a positive correlation between the number of these stress events experienced and the onset of illness (e.g. see Rahe et al., 1970). Even though we should be cautious about inferring causality here, because there are likely to be a myriad of additional factors which mediate this relationship (e.g. individual coping strategies), as we will see later in this chapter, the strength of social support received may help to "buffer" against some of the negative impacts of stressful life events, including illness. For this reason, increasing access to social support provision, whether this is from friends/family, community organisations, or government agencies would seem to be of paramount importance, particularly given the increasing demand on health services as the population continues to grow.

While most of us are able to call upon members of our support network in times of need, we know that there are certain individuals in society who, for various reasons, have limited access to social support. Additionally, there may be some people who will find it difficult to locate individuals within their offline networks who can relate to, empathise with, or provide valuable advice on their situation. This might be, for example, because they are living with an uncommon condition or illness or are experiencing problems that others cannot readily identify with. For these types of individuals, the Internet may be a particularly valuable resource to gain support that may otherwise be exceptionally difficult to locate. Knowing that more and more of us are turning to the Internet in times of need, it is imperative that we evaluate the efficacy of support offered in this environment and consider how we might maximise social support delivery online. Therefore, this chapter will address a number of key questions: 1) What are the major benefits of offering and receiving social support online? 2) Are there are any unique elements of the online world which make providing support more problematic? 3) How can psychological theory inform the design of support websites to improve the user experience? Before we do this, however, we will begin by defining social support and considering why it is so valuable to the individual.

7.2 What is social support and why is it important?

Social support can be broadly defined as a "transactional communicative process, including verbal and/or nonverbal communication, that aims to improve an individual's feelings of coping, competence, belonging, and/or esteem" (Mattson & Hall, 2011, p. 184). This definition acknowledges that support can be offered in a variety of different ways. For example, if your friend disclosed to you that he/she was considering splitting from their husband/wife, you might offer support in various capacities. A simple hug might communicate that you are sympathetic towards their plight and are there for them on an emotional level, and this would send a powerful message to your friend that he/she is not alone. On the other hand, whereas recommending a good solicitor in the event of a divorce might not help your friend to cope with the emotional aftermath of the split, it would be supporting them in a more practical sense by helping them to better understand the legal processes involved. This definition is also a good one because it recognises that there is potentially an assortment of benefits for the individual who is receiving the support, for example becoming better informed or increasing self-confidence and competence in dealing with the problems faced.

It is also worthwhile at this juncture differentiating between *actual* and *perceived* support. Whereas actual support refers to concrete examples of things that others might do for us in times of need (e.g. offering practical advice, talking through our issues with us etc.), perceived support refers to the support-seeker's perceptions regarding the availability and usefulness of the support provided (Mattson & Hall, 2011; Sarason et al., 1990). It is important to make this distinction because the two can be mutually exclusive. For example, an individual, on the surface, may receive abundant support from his/her network, but importantly might not perceive this support to be adequate or effective for their specific needs. For example, if you were living with a stigmatised illness, let's say depression, although your friends might offer you sympathy and words of encouragement, at the same time they might not fully understand what you are going through because they have not experienced depression personally. Of course, there is also the possibility that although members of your network might attempt to support you through your difficult patches, their naivety about the issues you face might also mean that even with the best of intentions they cannot offer you the type of support that you really desire. In some cases they might even make matters worse by advising you inappropriately or frustrating you by over-simplifying your problems or demonstrating their ignorance (e.g. by suggesting that you should "just get over it"). Sometimes there are instances when only someone who has walked in our shoes can truly appreciate what we are going through.

Schaefer et al. (1981) distinguish between five types of social support: emotional, esteem, network, information, and tangible. Emotional support refers to any type of supportive behaviour which attempts to impact positively on another individual's mood or emotional state. For example, this might include physical contact (e.g. a hug), messages of support, or even telling a joke to lift someone's spirits. Esteem support is concerned with attempts to strengthen another individual's self-worth or trying to instil in them the belief that they can face their problems and prosper. Network support relates to supportive behaviour which affirms that the person in need is not facing their problems alone, and that others are there for them when required. Information support consists of the communication of beneficial or necessary information. For example, this could include making a person aware of the types of options that are available to treat their specific illness or discussing with them what they might expect from a certain medical procedure. Finally, tangible support includes access to material resources (e.g. letting someone sleep in your spare room), physical assistance (e.g. driving someone to the hospital for an appointment), and financial assistance (House, 1981; Mattson & Hall, 2011; Schaefer et al., 1981). Clearly these different types of support are not necessarily mutually exclusive, and members of any social support network can offer multiple types of support simultaneously. Notably, it is possible for all of these types of support to be offered online, except for most aspects of tangible support, as this is the only form which is primarily communicated by physical actions rather than words.

An abundance of research evidence points to the notion that increased social support has been linked with improvements in both physical and mental health (e.g. see Broadhead et al., 1983; Coker et al., 2002; House et al., 2001; Lyyra & Heikkinen, 2006). There are a number of theories to explain this effect; including the often cited *stress-buffering hypothesis* (Cobb, 1976; Cohen & Wills, 1985). Put simply, the perception that others are responsive to our needs and able to support us through difficult times may strengthen our own resolve and perceived capacity to deal with stresses, therefore leading to a reappraisal of the stressful event(s). In essence, the supported individual will feel more able to cope with any demands that are placed on him/her and therefore this will lessen or eradicate the stress reaction (Cohen & Wills, 1985). Because we know that there are physiological symptoms associated with stress (e.g. heart disease, disturbed sleep patterns) and that people experiencing stress are likely to adopt more unhealthy habits (e.g. taking drugs, misusing alcohol, over-eating), the negative health consequences associated with these symptoms and behaviours may be minimised when effective social support is in place (Cohen & Wills, 1985; Mattson & Hall, 2011). Fundamental to this theory is also the idea that the benefits of social support are only really evident when a person is experiencing high levels of stress (Cohen & Wills, 1985). This is in contrast to the *direct effects hypothesis*, which predicts that social support will provide continuous benefits. There is evidence in the experimental literature to support both models (see Cohen & Wills, 1985), so consensus has yet to be reached as to which hypothesis is the most accurate. However, irrespective of which of these theories has the most predictive power, the common link between them is that receiving social support from others comes with a host of physiological and psychological benefits for the individual. Of course, there are also many other potential explanations for this relationship. For example, with a larger support network, it is likely that individuals will receive increased amounts of tangible and information support which will not only help them to stay healthy (e.g. offering advice on healthy living, and making sure that appointments are met and medication is taken) but also to recover from their illness (e.g. assisting with basic needs so the person can recuperate) (Mattson & Hall, 2011). So, we have a relatively clear idea about how we can support people in our offline lives, but how might we support people in the online world?

7.3   How is social support offered online?

Online support groups can be defined as a "type of virtual community with a health-related focus, which provide an online environment where individuals can connect and interact with other people who have had similar experiences to exchange information, social support or advice" (Coulson & Smedley, 2015, p. 198). Online support groups are typically set up to support individuals with particular medical conditions (e.g. diabetes, cancer, narcolepsy) and therefore communication normally centres on the condition itself, but can also extend to other topics, including relationships and politics (Coulson & Smedley, 2015; Finn, 1999). Online communities are not, however, restricted to a health focus, and there are innumerable topics that cover the gamut of human experience and the problems and issues that we all face in our everyday lives. A brief search of dailystrength.org, for example, uncovered support groups on topics as diverse as "acne," "body modification," "divorce," "bullying," "empty nests," "pet bereavement," and "coming out." Individuals can communicate with one another in various different ways via support communities; however, forums tend to be the most common mode of interaction (Coulson & Smedley, 2015). Forums have many distinct advantages over other forms of online communication (e.g. chat rooms).

Most notably, because communication is asynchronous in nature (i.e. it does not take place in real time), members can post messages and respond to others at their own convenience and do not need to be online at the same time as the individual(s) with whom they are interacting (Barak et al., 2008; Coulson & Smedley, 2015). Additionally, the fact that messages are archived in different threads or topics means that members can benefit from the responses given to other individuals, in addition to those which are directed solely at them. As previously noted, it is not possible to interact physically with someone else (e.g. hug them) online but all of the other types of social support outlined by Schaefer et al. (1981) can be offered in an online setting.

7.4 Why might people seek support online instead of offline?

Although the range of topics that online support communities cover are diverse, scholars suggest that there are likely to be specific socio-psychological characteristics which predict one's inclination to join an online support group. For example, it has been argued that finding satisfactory support in the offline world might be difficult if the individual is living with a condition which is rare (i.e. because it would be hard to find others who could relate to them), hidden (i.e. because others might not appreciate the full impact that it has on the person), or not valued highly by society (i.e. because people might not see the condition as having genuine health repercussions) (McKenna & Bargh, 1998; Mickelson, 1997). Individuals who fall into these specific groups may be particularly motivated to seek support from others online (Cummings et al., 2002). If individuals perceive support in their offline networks to be inadequate, but support in their online communities as satisfactory, then they are likely to spend more time engaging with online support groups (Turner et al., 2001). In essence, people turn to online support communities when they feel that they cannot receive the depth of support that they require in the offline world, however, they will only continue to participate if their online support group can offer them the quality of support that they desire. Turner et al. (2001) also provided support for Cutrona and Russell's (1990) "optimal matching" hypothesis, which proposes that different types of supportive behaviour are required under different types of stress events. Moreover, they assert that the extent to which the individual has control over the stress event(s) plays an important role in dictating their social support requirements. In situations where the person has little control over their circumstances, for example the sudden onset of a medical illness, a strong need for emotional support is created, partly because this generates a sense of helplessness in the individual. When these emotional needs cannot be met via offline networks, the person may seek support elsewhere, for instance in the online world. Because we know that for certain types of individuals online communities will be their primary source of social support, it would seem necessary to evaluate whether satisfactory support can be provided in this context. We will now therefore turn our attention to the advantages and disadvantages of providing support online.

7.5   Advantages of online support

There are potentially numerous advantages to offering and receiving support in an online environment. At a very basic level, unlike offline support groups, the individual would not need to travel to the support venue and could access the group at his/her own convenience and in the comfort of their own home (Turner et al., 2001). Clearly this has the potential to save the person time and money and may present a more relaxing and comfortable environment in which to converse with others, which may be attractive to those who are shy or who find it difficult, stressful or embarrassing to communicate face-to-face. Indeed, cyber-psychological theory suggests that we have far more control over our presentation of self in cyberspace. For example, Walther's "hyperpersonal" theory of communication argues that asynchronous forms of communication may lead to optimal self-presentation because, for example, the individual can more carefully consider the messages they write before posting them (Walther, 1996; Walther and Parks, 2002). Additionally, this should be particularly advantageous to individuals who live in geographically remote areas or who may have difficulties leaving the house (e.g. individuals with physical disabilities or people who suffer from agoraphobia). Because online groups are not restricted in the same way as offline groups in terms of space, group sizes can be much larger. Individuals from all corners of the planet could potentially contribute in an online community, meaning that members will be exposed to a wide array of different outlooks, experiences and ideas (White & Dorman, 2001). Additionally, online interaction should be more socially liberating and egalitarian because individuals can communicate with less inhibition. The online world can be a great leveller because many of the cues that we might traditionally associate status and authority (e.g. body language) are missing or attenuated, meaning that, theoretically at least, people can interact on a level playing field (Fullwood et al., 2011; Suler, 2004). Individuals can also choose to hide specific characteristics (e.g. ethnicity, disability, sexuality, sex) if they feel that there is a potential that they may be discriminated against or excluded from interactions (White & Dorman, 2001).

One of the major advantages of online support groups comes with the freedom that users have to choose which aspects of their identity they reveal. A member of an online support community may choose to remain completely anonymous if he/she wishes to do so. In addition, members will communicate predominantly with strangers or individuals whom they have never met face-to-face. Therefore, even when some identity information is disclosed, individuals should still feel a sense of protection and safety because it is highly unlikely that they will bump into fellow support members in their offline lives by mere chance (Fullwood et al., 2013). An upshot of increased perceived anonymity is that some members (particularly those who live with stigmatised conditions) may feel more comfortable discussing difficult, sensitive, or potentially embarrassing issues, partly because the level of risk is reduced (Adelman et al., 1987; White & Dorman, 2001). This is important because a cognitive reappraisal of a difficult emotion or distressing experience (which may help to reduce emotional distress) is more likely to take place when effective comforting communication is given. Moreover, it is argued that a number of conditions need to be met for effective comforting communication to occur; namely: self-disclosure, discussion concerning thoughts and feelings and discussion concerning reappraisals (Burleson & Goldsmith, 1998). In essence, a person will receive more effective emotional support if they are willing to talk honestly and in detail about their problems and emotional issues in an environment which fosters reappraisals (i.e. one which allows them to work through the difficulties that they are facing). Fullwood and Wootton (2009) provided some evidence to suggest that in an online forum dedicated to epilepsy, members who were anonymous spent more time discussing their thoughts and feelings about upsetting experiences and emotions than members who identified themselves. On the surface this suggests that anonymity helps people to feel more comfortable about opening up and overcoming the self-presentation dilemma that people often face when discussing sensitive and personal issues face-to-face (Burleson & Goldsmith, 1998; Caplan & Turner, 2007). Finally, communicating primarily with strangers may mean that members receive a much more objective and honest appraisal of their situation than they might get from someone who is more emotionally involved, for example a friend or family member (Adelman et al., 1987; Walther & Boyd, 2002).

7.6   Disadvantages of online support

Perhaps the biggest drawback to offering support online is the assumption that the individual in need of support will actually have access to the Internet and/or is capable of navigating the World Wide Web effectively. We know, for example, that a digital divide still exists for many elements of society, for example those with learning difficulties are still less likely to enter the online world (see e.g. Chadwick et al., 2013 and Chapter 1). The unfortunate irony is that these groups of individuals often stand to gain the most (e.g. developmentally, socially) from being online. For those who can get online, being a member of an online support community is not unproblematic. Moderators may be stretched to capacity, particularly in larger groups, and may not always be able to monitor and shut down dangerous, hostile or dishonest communications. In any online community, misinformation or unsuitable advice is always a possibility, particularly from members who are not medical experts. Clearly this has the potential to pose risks to the individual who might follow this advice unwittingly (Culver et al., 1997; Dickerson et al., 2000). Online communities may also attract people who actively want to abuse and attack other members and remain undetected whilst doing so (White & Dorman, 2001). Even those who do not have this aim may find themselves being drawn into hostile interactions because people are less inhibited online (see Suler, 2004). Moreover, these types of negative experiences are likely to discourage people from returning.

Interacting with others online is also likely to have a number of impacts on communication dynamics. There has been some suggestion for example that online messages can be more difficult to interpret because of the lack of contextual cues (White & Dorman, 2001). We know that paralinguistic features of speech are particularly important in terms of communicating emotion or affect. If we take sarcasm for instance, this is communicated almost entirely with tone and intonation, which would be absent from text-based communication. Group members may also find it more difficult to express emotions because they cannot make use of nonverbal communication (e.g. facial expressions, touch, gestures). Although this can be compensated for to some extent by the use of paralinguistic features of text, for example emoticons, capitalisation, and multiple punctuation (Crystal, 2001; White & Dorman, 2001), interacting online may encourage a more loosely-structured and informal pattern of communication, including the use of more slang and textspeak (e.g. acronyms like "LOL" for "Laugh Out Loud") (Scott et al., 2014). Moreover, some aspects of textspeak may result in less favourable impressions of those who use them (Fullwood et al., 2015), however the implications of how this might affect one's inclination to offer support to others has not been investigated.

Whereas, in face-to-face groups, participation is encouraged and often mandatory, members of online groups can "lurk" in the background, only reading others' messages and without making any real contribution to the community (Dickerson et al., 2000; White & Dorman, 2001). Clearly, this has implications for the practicability of a group to continue if more members lurk than participate. There are likely to be a number of explanations for why people lurk, for example not understanding how to use the site, not feeling that they can make a valuable contribution or not feeling that their needs can be met (Preece et al., 2004). Further, lurkers feel that they receive less social support than those who do contribute and are less satisfied with their relationships with other members of the group (Mo & Coulson, 2010), so more should be done to encourage their participation (and this will be addressed in the next section).

Of those who do participate, there may also be an issue regarding whether their contributions are beneficial to others in the community or whether they use the site purely to meet their own needs. Clearly an online group can only function if members offer support in addition to seeking it. Research by Venner et al. (2012) suggests that anonymity may play a key role in determining the type of participation that members make. Drawing on SIDE theory (Social Identity of Deindividuation Effects) (Reicher et al., 1995; Spears & Lea, 1992), their research suggests that the balance of anonymous vs. identifiable individuals in a support community is likely to have a strong impact on participation activity. When a group contains a mixture of anonymous and identifiable members, those who are anonymous will be more likely to use the group to meet their own needs (i.e. by asking for help) and less likely to support others. This has implications for group cohesion and whether or not support groups should encourage members to identify themselves, and this will be discussed later.

7.7   Design implications and enhancing the user experience

In attempting to improve the user experience, perhaps the most obvious issue to address relates to the non-participation of lurkers, particularly given estimates which suggest that lurking rates can range anywhere from 45.5% to 90% of an entire community (e.g. see Mason, 1999; Nonnecke, 2000; Nonnecke & Preece, 2000). Lurking may be problematic if it means that very few people are making contributions, and members are likely to turn to other groups when they do not receive responses to their posts. Further, online communities will thrive from the richness that is provided by more members sharing different experiences, ideas, and perspectives. Therefore, turning lurkers into active participators should be one key goal of any online support community. Preece et al.'s (2004) research into the reasons why people lurk on online support groups is not only enlightening because it suggests that lurkers are not always "selfish free riders" (p. 221), but also because it offers a number of tips to site developers and moderators to encourage participation. For example, a clear policy statement (preferably communicated somewhere visible, e.g. in a welcome statement) should be included on the site, out-lining how contributions from all members are welcome, and vital for the group to function. Moderators may also take the lead on personally inviting lurkers to get involved, and may use tracking tools to monitor contributions. Finally, they suggest that online communities should take a leaf out of the book of e-commerce sites by rewarding contributors for the quality and quantity of contributions that they make. Although it may not be feasible to offer financial incentives, rewards may be offered in other ways. For example, the site could publish a list of "top con-tributors" and update this regularly. The powerful influence of reward on reinforcing behaviours has been well documented in the psycho-logical literature (e.g. see Skinner, 1938). Additionally, members could be encouraged to rate and leave reviews on other contributors, much in the same way that customers leave feedback for sellers on Amazon for example. At the same time, however, the authors acknowledge that close monitoring of activities would be necessary to ensure that competitors do not purposefully leave negative reviews to bolster their own status in the community or that members do not simply get their friends to write false positive reviews about them.

Another challenge faced by an online support group is in how to attract new members to their community. New members are always needed to substitute for those who have left the site. In addition, new members can help to keep the community fresh by offering new pers-pectives and ways of thinking about problems (Ren et al., 2007). In the first instance, a search engine optimisation strategy should be in place (for example, ensuring that relevant keywords are included in the con-tent of the pages) to help increase traffic to the site. Strategies may also be used to transform visitors into fully-fledged members. For example,

Ren et al. (2007) suggest lowering barriers to entry (e.g. only asking for necessary information upon joining), actively encouraging members to post once they have signed up, and providing rapid feedback to any questions that are asked early in the membership cycle. Although a support community should take all steps necessary to recruit new members, this should be balanced against the potential that this might upset or irritate more established members. For example, new members will likely pose questions that longer-term members would have seen before on countless occasions. They may also be more likely to disregard or unintentionally violate the rules and conventions of the community, partly due to a lack of experience but also owing to the fact that they haven't built up the same level of emotional investment in the site (Ren et al., 2007). Certain approaches may therefore be employed to ensure that newcomers do not disrupt group cohesion. For example, the site would benefit from a frequently asked questions section, as well as a highly visible mission statement and "rules of conduct." Additionally, new members could begin their journey in a "newbie garden," or be paired up with a more regular user of the site for the purpose of mentorship (Ren et al., 2007).

One of the biggest challenges faced by online support communities is in promoting and maintaining a collective identity. It would be beneficial for members to feel as if they are part of the community and have a strong affiliation to the group; particularly given that one's social identity (i.e. the extent to which our self-concept is derived from our membership to specific groups) (Tajfel & Turner, 1979; Turner, 1985) has been shown to correlate positively with perceived social support and life satisfaction (Haslam et al., 2005). In essence, members will be more likely to remain loyal to the group if they feel a strong connection to it and a sense of shared identity provides a strong foundation for both providing and receiving support (Haslam et al., 2004, 2005; Levine et al., 2002, 2005). Building and maintaining a collective identity might be hard to achieve as a community expands and as individual members' contributions get lost amongst the multitude of posts. One way to overcome this might be to break down the larger community into smaller "neighbourhoods," and this might be based on geographical location, for example (Ren et al., 2007). In addition, it is likely that members will create more meaningful bonds with other users if they actively engage in both helping and seeking behaviours. Finding ways to encourage participation on both fronts would therefore likely pay dividends. One possible method to achieve this would be to create both "anonymous" and "identified" neighbourhoods. Theoretically, it should be easier to create a sense of shared purpose in a community if all members are on a level playing field and, as previously noted, a group which contains a mixture of anonymous and identified members tends to result in the anonymous individuals using the group in a more selfish fashion (Venner et al., 2012). Additionally, online support communities should encourage users to identify themselves where possible. Clearly, however, this must be balanced against the potential benefits associated with remaining anonymous to other users, and might not be feasible for all types of communities. Encouraging members to offer support as well as seek it will not only help promote a collective identity, but should also produce a variety of other benefits. For example, research has shown that in helping others, we might increase our own self-worth, and sense of purpose and belonging (see e.g. Taylor & Turner, 2001).

Finally, a number of additional strategies may be undertaken to improve the user experience. Communities need a strong presence from moderators who should be selected carefully on the basis of their dedication and subject-specific knowledge. Moderators should play a key role in monitoring illegal, anti-social, and inappropriate conduct. A clear policy on dealing with aggressive and anti-social users should be in place. For example, persistent offenders should have their membership revoked to maintain a sense of harmony in the community. It is also important that codes of conduct are made clear, so that no doubt will be left in the minds of members as to what is acceptable behaviour and what is not. Communities may also consider requiring their members to log in to the site and ensuring that contributions to the forum are password protected and not in the public domain. This will create a sense of protection and may encourage more members to reveal their identities, while at the same time giving them a sense of freedom to self-disclose. In addition, this would likely discourage those whose sole purpose would be to harass and attack other members, as signing up would entail some effort. Furthermore, considering the limitations that text-only communication places on our ability to express ourselves (particularly on an emotional level), sites could incorporate elaborate emoticon creation facilities and the ability to communicate via other modalities (e.g. video chat) if required. Lastly, some consideration should be given to making online communities as inclusive and accessible to all as possible. This should include adherence to Universal Design principles to ensure that people with disabilities can make effective use of the support community (see Chapter 1 for wider coverage of some of these issues).

7.8 Summary and conclusion

In this chapter we have noted that for some individuals, joining online communities may be the most effective option available to have their support needs met. This might be, for example, because the condition that they are living with is uncommon and finding others who understand their experiences offline would not be easy. We have also discussed the many benefits that come with providing and seeking support in cyberspace. For example, members have more control over the amount of identity information that they disclose which gives them a greater sense of freedom and protection to discuss sensitive, embarrassing, and private topics. As well as there being a host of advantages, we have also outlined a number of shortcomings associated with online support groups. For instance, the distanced and often anonymous nature of interaction means that members are under no obligation to participate. There are a number of challenges faced by online support communities (e.g. attracting new members, dealing with lurkers); however, within this chapter a variety of strategies have been proposed, grounded in research evidence and theory, which may be employed to enhance the user experience. Adopting some or all of these strategies should produce a more comfortable, safe, and inclusive environment in which to discuss experiences with others, leading to more effective social support provision.

7.9 References

Adelman, M. B., Parks, M. R., & Albrecht, T. L. (1987). Beyond close relationships: Support in weak ties. In T. L. Albrecht & M. B. Adelman (Eds.), *Communicating social support* (pp. 126–147). Newbury Park, CA: Sage.

Barak, A., Boniel-Nissim, M., & Suler, J. (2008). Fostering empowerment in online support groups. *Computers in Human Behavior, 24*(5), 1867–1883.

Broadhead, W. E., Kaplan, B. H., James, S. A., Wagner, E. H., Schoenbach, V. J., Grimson, R., Heyden, S., Tibblin, G., & Gehlbach, S. H. (1983). The epidemiologic evidence for a relationship between social support and health. *American Journal of Epidemiology, 117*(5), 521–537.

Burleson, B. R., & Goldsmith, D. J. (1998). How the comforting process works: Alleviating emotional distress through conversationally induced reappraisals. In P. A. Anderson & L. K. Guerrero (Eds.), *Handbook of communication and emotion: Theory, research, application, and contexts* (pp. 245–280). San Diego, CA: Academic Press.

Caplan, S. E., & Turner, J. S. (2007). Bringing theory to research on computer-mediated comforting communication. *Computers in Human Behavior, 23*(2), 985–998.

Chadwick, D., Wesson, C., & Fullwood, C. (2013). Internet access by people with intellectual disabilities: Inequalities and opportunities. *Future Internet, 5*(3), 376–397.

Cobb, S. (1976). Social support as a moderator of life stress. *Psychosomatic Medicine, 38*(5), 300–314.

Cohen, S., &Wills, T. A. (1985). Stress, social support and the buffering hypothesis. *Psychological Bulletin, 98*(2), 310–357.

Coker, A. L., Smith, P. H., Thompson, M. P., McKeown, R. E., Bethea, L., & Davis, K. E. (2002). Social support protects against the negative effects of partner violence on mental health. *Journal of Women's Health and Gender-Based Medicine, 11*(5), 465–476.

Coulson, N., & Smedley, R. (2015). A focus on online support. In A. Attrill (Ed.), *Cyberpsychology* (pp. 197–213). Oxford: Oxford University Press.

Crystal, D. 2001. *Language and the Internet.* Cambridge University Press: Cambridge.

Culver, J. D., Gerr, F., & Frumkin, H. (1997). Medical information on the Internet: A study of an electronic bulletin board. *Journal of General Internal Medicine, 12*(8), 466–470.

Cummings, J. N., Sproull, L., & Kiesler, S. B. (2002). Beyond hearing: Where real-world and online support meet. *Group Dynamics: Theory, Research and Practice, 6*(1), 78–88.

Cutrona, C., & Russell, D. (1990). Type of social support and specific stress: Toward a theory of optimal matching. In B. Sarason, I. Sarason, & G. Pierce (Eds.), *Social support: An interactional perspective* (pp. 319–366). New York: Wiley.

Dickerson, S. S., Flaig, D. M., & Kennedy, M. C. (2000). Therapeutic connection: Help seeking on the Internet for persons with implantable cardioverter defibrillators. *Heart and Lung, 29*(4), 248–255.

Donne, J. (1999). *Devotions upon emergent occassions/Death's duel.* Random House: New York.

Finn, J. (1999). An exploration of helping processes in an online self-help group focusing on issues of disability. *Health and Social Work, 24*(3), 220–231.

Fullwood, C., Evans, L., & Morris, N. (2011). Linguistic androgyny on MySpace. *Journal of Language and Social Psychology, 30*(1), 114–124.

Fullwood, C., Melrose, K., Morris, N., & Floyd, S. (2013). Sex, blogs and baring your soul: Factors influencing UK blogging strategies. *Journal of the American Society for Information Science and Technology, 64*(2), 345–355.

Fullwood, C., Quinn, S., Chen-Wilson, J., Chadwick. D., & Reynolds, K. (2015). Put on a smiley face: Textspeak and personality perceptions. *Cyberpsychology, Behavior and Social Networking, 18*(3), 147–151.

Fullwood, C., & Wootton, N. (2009). Comforting communication in an online epilepsy forum. *Journal of Cybertherapy and Rehabilitation, 2*(2), 159–164.

Haslam, S. A., Jetten, J., O'Brien, A., & Jacobs, E. (2004). Social identity, social influence, and reactions to potentially stressful tasks: Support for the self-categorization model of stress. *Stress and Health, 20*(1), 3–9.

Haslam, S. A., O'Brien, A., Jetten, J., Vormedal, K., & Penna, S. (2005). Taking the strain: Social identity, social support and the experience of stress. *British Journal of Social Psychology, 44*(3), 355–370.

Holmes, T. H., & Rahe, R. H. (1967). The social readjustment rating scale. *Journal of Psychosomatic Research, 11*(2), 213–218.

House, J. S. (1981). *Work stress and social support.* Reading, MA: Addison-Wesley.

House, J. S., Landis, K. R., & Umberson, D. (2001) Social relationships and health. In Conrad, P. (Ed.), *The sociology of health and illness: Critical perspectives.* New York: Worth Publishers.

Levine, R. M., Cassidy, C., Brazier, G., & Reicher, S. D. (2002). Self-categorization and bystander non-intervention: Two experimental studies. *Journal of Applied Social Psychology, 32*(7), 1452–1463.

Levine, R. M., Prosser, A., Evans, D., & Reicher, S. D. (2005). Identity and emergency intervention: How social group membership and inclusiveness of group boundaries shape helping behavior. *Personality and Social Psychology Bulletin, 31*(4), 443–453.

Lyyra, T. M., & Heikkinen, R. L. (2006). Perceived social support and mortality in older people. *The Journals of Gerontology Series B: Psychological Sciences and Social Sciences, 61B*(3), S147–S152.

Mason, B. (1999). Issues in virtual ethnography. *Paper Presented at the Ethnographic studies in real and virtual environments: Inhabited information spaces and connected communities conference*, Edinburgh.

Mattson, M., & Hall, J. G. (2011). *Health as a communication nexus: A service-learning approach.* Iowa: Kendall Hunt Publishing Company.

McKenna, K., & Bargh, J. (1998). Coming out in the age of the Internet: 'Demarginalization' through virtual group participation. *Journal of Personality and Social Psychology, 75*(3), 681–694.

Mickelson, K. D. (1997). Seeking social support: Parents in electronic support groups. In S. Kiesler (Ed.), *Culture of the Internet* (pp. 157–178). Mahwah, NJ: Erlbaum.

Mo, P. K. H., & Coulson, N. S. (2010). Empowering processes in online support groups among people living with HIV/AIDS: A comparative analysis of 'lurkers' and 'posters'. *Computers in Human Behavior, 26*(5), 1183–1193.

Nonnecke, B. (2000). *Lurking in email-based discussion lists.* Unpublished PhD dissertation, South Bank University, London.

Nonnecke, B., & Preece, J. (2000). Lurker demographics: Counting the silent. *Paper Presented at the ACM CHI 2000 conference on human factors in computing systems*, The Hague.

Preece, J., Nonnecke, B., & Andrews, D. (2004). The top five reasons for lurking: Improving community experiences for everyone. *Computers in Human Behavior, 20*(2), 201–223.

Rahe, R. H, Mahan, J. L., & Arthur, R. J. (1970). Prediction of near-future health change from subjects' preceding life changes. *Journal of Psychosomatic Research, 14*(4), 401–406.

Reicher, S. D., Spears, R., & Postmes, T. (1995). A social identity model of deindividuation phenomena. *European Review of Social Psychology, 6*(1), 161–98.

Ren, Y., Kraut, R., & Kiesler, S. (2007). Applying common identity and bond theory to design of online communities. *Organization Studies, 28*(3), 377–408.

Sarason, B. R., Sarason, I. G., & Pierce, G. G. (1990). *Social support: An interactional view.* New York: John Wiley.

Schaefer, C., Coyne, J. C., & Lazarus, R. S. (1981). The health-related functions of social support. *Journal of Behaviorial Medicine, 4*(4), 381–406.

Scott, G. G., Sinclair, J., Short, E., & Bruce, G. (2014). It's not what you say, it's how you say it: Language use on Facebook impacts employability but not attractiveness. *Cyberpsychology, Behavior and Social Networking, 17*(8), 562–566.

Skinner, B. F. (1938). *The behavior of organisms: An experimental analysis.* New York: Appleton-Century.

Spears, R., & Lea, M. (1992). Social influence and the influence of the 'social' in computer-mediated communication. In M. Lea (Ed.), *Contexts of computer-mediated communication* (pp. 30–65). Hemel-Hempstead, UK: Harvester-Wheatsheaf.

Suler, J. (2004). The online disinhibition effect. *Cyberpsychology, Social Networking and Behavior, 7*(3), 321–326.

Tajfel, H., & Turner, J. C. (1979). An integrative theory of intergroup conflict. In W. G. Austin & S. Worchel (Eds.), *The social psychology of intergroup relations* (pp. 33–47). Monterey, CA: Brooks/Cole.

Taylor, J., & Turner, J. (2001). A longitudinal study of the role and significance of mattering to others for depressive symptoms. *Journal of Health and Social Behavior, 42*(3), 310–325.

Turner, J. C. (1985). Social categorization and the self-concept: A social cognitive theory of group behaviour. In E. J. Lawler (Ed.), *Advances in group processes* (Vol. 2, pp. 77–122). Greenwich, CT: JAI Press.

Turner, J. W., Grube, J. A., & Meyers, J. (2001). Developing an optimal match within online communities: An exploration of CMC support communities and traditional support. *Journal of Communication, 51*(2), 231–251.

Venner, R., Galbraith, N., & Fullwood, C. (2012). Anonymity and level of support given on a health-related online support forum. *Journal of Cybertherapy and Rehabilitation, 5*(1), 9–13.

Walther, J. B. (1996). Computer-mediated communication: Impersonal, interpersonal, and hyperpersonal interaction. *Communication Research, 23*(1), 3–43.

Walther, J. B., & Boyd, S. (2002). Attraction to computer-mediated support. In C. A. Lin & D. Atkin (Eds.), *Communication technology and society: Audience adoption and uses of the new media* (pp. 133–167). New York: Hampton Press.

Walther, J. B., & Parks, M. (2002). Cues filtered out, cues filtered in. In M. L. Knapp & J. A. Daly (Eds.), *Handbook of interpersonal communication* (pp. 529–563). Thousand Oaks, CA: Sage.

White, M., & Dorman, S. M. (2001). Receiving social support online: Implications for health education. *Health Education Research: Theory and Practice, 16*(6), 693–707.

Withers, B. (1972). *Lean on me*, on 'Still Bill' (CD). The Record Plant, Los Angeles: Sussex Records.

8
Counselling in Online Environments

Rachel Harrad and Nick Banks

8.1 Introduction

The Internet is no longer regarded as a world separate to our physical existence, but rather exists more as part of individuals' everyday lives as a method to interact with others (Wellman & Haythornthwaite, 2002). The Internet has changed the way individuals obtain information, explore, and expand relationships with others in ways not contemplated until relatively recently (Skinner & Zack, 2004). It is being explored for both legitimate professional and charlatan therapies by practitioners, potential practitioners, clients, and conmen alike. One feature of the Internet that has developed in recent times is that of online counselling (Mallen et al., 2011). It is therefore important to consider academic theory and research in relation to online counselling methods to help further our understanding of the possible effectiveness of online therapies, and the factors that impact this effectiveness in order for clients to gain the best clinical outcomes.

Within the United Kingdom (UK), both professional counselling and psychotherapy have increased their online presence (Fletcher-Tomenius et al., 2009) and many have stressed that the nature of online counselling enhances and expands opportunities to connect people, rather than being just an alternative or substitute for face to face (FtF) interaction (Skinner & Zack, 2004). It is with this in mind that this chapter begins by considering traditional (offline) counselling compared to online counselling, prior to evaluating online counselling, from a cyberpsychological perspective, to offer recommendations for practice.

8.2   Approaches in online counselling

The Internet has provided a different and diverse medium for the development of therapeutic working relationships for counselling. Similarly to traditional counselling methods, a variety of terms exist (Chester & Glass, 2006), including *Internet therapy* (Rochlen et al., 2004) or *Internet counselling* (Pollock, 2006) or even E-Therapy. Various definitions of online therapy exist, including any form of therapeutic intervention that uses the Internet to deliver the service between professional and client (Rochlen et al., 2004). For the purpose of this chapter, we will use the terms *online therapy* and *online counselling* interchangeably to refer to text-based modalities conducted at a distance, as well as interventions conducted via video-link. Whilst both Internet counselling and traditional counselling methods struggle with clear definition, the Internet builds on traditional counselling methods. In the past, these have included therapy based on telephone or letter interaction (Rochlen et al., 2004) as well as FtF interaction. Using electronic communications is therefore not new. They have been historically used for therapeutic interventions (Skinner & Zack, 2004) and, similarly to more traditional types of therapy, the communications offered by the Internet can be synchronous or asynchronous. *Synchronous communications* involve interaction between client and therapist at the same point in time, whilst *asynchronous communications* operate at different points in time, where client and therapist respond when able (Suler, 2000). Online therapy programmes incorporate both asynchronous and synchronous communications, such as psychological interventions via email, or online counselling in real time, for instance via chat rooms, self-help groups, or online health information (Proudfoot, 2004).

8.2.1   The time-sensitive approach

Online counselling techniques employed via these mediums tend to be developed from what Jones and Stokes (2008) refer to as a *time-sensitive approach*, whereby counsellors working online should develop conversations that have richness and depth rather than long duration. In line with this view, Bor et al. (2004) suggest that online counselling approaches tend to lend themselves more to brief, time-sensitive approaches. In contrast, therapists working in a FtF situation may not feel that they have to work to a brief fixed schedule. Emails from a client may be long or short, depending on how they wish to express and describe their needs, and can be sent without a clear need to hold on to information until the next session. This may require some ground rules to be put in place within a counselling contract of expectations, so that clients do not feel they are ignored when emails are not immediately responded to. It may be that with some clients, the counsellor needs to work at increasing the depth of the written communication from the client.

Whilst there are a variety of different asynchronous and synchronous methods of therapy operating within the online domain, this chapter is limited to those where therapist and client are in different geographical locations. This chapter does not intend to be an exhaustive review of the many types of online therapies available. Moreover, it attempts to consider some of the themes identified within popular online counselling modalities. The modalities considered in this chapter are *therapy via electronic mail* or email, *therapy via online chat* or instant messaging, and *communications via webcam* or videoconferencing in a one-to-one format between client and therapist. It is noted that therapy might also ensue in a group situation, online or offline, but for this book, we have chosen to focus on the one-to-one therapeutic relationship. Subsequently, brief attention will be given to the different types of therapy that might ensue via these modes of communication online.

8.3   E-mail therapy

Electronic mail or email offers an easy and convenient way for individuals to communicate using the Internet (Goss & Anthony, 2003). Email counselling is an asynchronous method where client and therapist interact within a time of their own choosing, and correspond via emails. This negates the need for a formal appointment, as is the case with traditional therapy (Suler, 2000) and other online methods, such as instant chat or webcam (Goss & Anthony, 2003). Individuals do not need to distract themselves with each other's physical appearance or how their tone of voice projects to others. They are also able to avoid potentially off-putting cues from the therapist, such as reactions to disclosure (Maples & Han, 2008). As such, clients may be more open to disclosing distressing and painful information to their therapist (Goss & Anthony, 2003).

Research considering the experiences of online counsellors has suggested that clients may be used to *talking in their heads* and working via email may represent an extension of this process. You might consider how you deal with your own issues or problems. The colloquial term for this process might be to mull something over. We have also often heard the addage that a problem shared is a problem solved. Email might offer alternative ways of carrying out these processes by offering a *half-way house* between reflection and sharing with others (Dunn, 2012). You can process your feelings in a reflective manner by writing them down, reading, editing, re-reading, re-editing etc., until you feel that you have got the content right in terms of reflecting what you really want to say to the therapist. This may be understood via the cyberpsychological concepts of *solipsistic Introjection* and *dissociative Imagination* (Suler, 2004). Suler (2004) suggests that within online communications it is possible to perceive conversations with others as taking place within the mind of the individual (*solipsistic Introjection*), which permits an individual to be more open, whilst it is also possible to perceive online interactions as something different and less real than a FtF exchange (*dissociative Imagination*).

Unhindered by the therapeutic hour of traditional therapy, issues can be explored at the client's pace via email (Goss & Anthony, 2003), with the increased ability to edit, delete, and modify one's self presentation. The *hyperpersonal model* (Walther, 1996) may offer assistance in interpreting these concepts. This model proposes that features within computer-mediated communications may enable optimal self-presentation. Specifically, individuals can selectively present the self to others, capitalising on the asynchronous nature of many online spaces, and the space afforded to ensure that they edit and re-edit their presentation where necessary to explain themselves fully in their desired way. More recent research in this area has suggested, however, that additional factors are relevant within computer-mediated communications, including the motivations of the individual's interaction partners (Walther, 2007). For example, within an email exchange, clients may anticipate further interactions with the therapist, and therefore may present in a more considered way than if they are engaging in an alternative modality, such as a single session of online chat. There is no denying, though, that counselling via email does offer increased opportunity for reflection between sessions.

Despite being an asynchronous method, email exchange offers the advantage of immediacy, with a client being able to write an email to the therapist at the moment in time a problem occurs. It has been suggested that often by the time a client arrives at a therapy session in FtF settings, any troubling feelings may have passed, and clients are uncertain of the nature of the upset (Murphy & Mitchell, 1998). Writing down feelings in the moment permits an *emotional logging* of the difficulties. A similar process is sometimes used in FtF counselling, such as in reflective logs (Wright & Bolton, 2012), with good therapeutic process outcome. The difference between the online and FtF process may be the degree to which it is shared with the counsellor, although this may reflect personal preference. Further, unlike FtF counselling, whereby an individual is required to reflect on a session based on memory/selective memory, email exchanges permit the storage of a permanent record of conversational exchange between client and therapist, which can be re-examined at later dates as part of a reflective process (Suler, 2000). Interviews with clients have revealed that clients cite benefit in the permanent record permitted by an email, which enables them to get more out of the counselling experience. Counsellors have also reported that they perceived this to be a beneficial aspect of online counselling (Dunn, 2012). Chechele and Stofle (2003) cite a client who found, on returning to an email exchange after time away, that the email was more helpful than it had been on initial reading, and enabled a different interpretation. However, there is limited research into how clients should be guided in this activity to ensure best clinical outcomes. Other issues may also arise surrounding the ambiguity of text in this method.

Ambiguities may not be clarified or explored as easily as in FtF counselling. When counselling via email the therapist might not have the visual clues to hand that s/he normally uses to detect various emotions, understandings or misunderstandings in the client in a FtF interaction. In FtF counselling, for instance, a therapist might immediately pick up on cues of a client misunderstanding or misinterpreting their intended communication. If this does not happen in an email exchange, the client might feel emotionally wounded and withdraw from seeking further clarification. Of particular note is Jones and Stokes (2008) commentary on the increased potential for unsettling experiences through the psychodynamic theory processes of *transference, counter-transference,* and *projection,* which they suggest may become more powerful when presented through typed communication. For example, in considering the process of *transference,* if there is a technological breakdown on the part of client or therapist it may be useful to explore the impact of this on the client and the level at which they express feelings of hurt, abandonment or rejection in the context of the *paused communication* being related to technical issues entirely beyond the therapist's or client's control. There are many factors that come into play in these situations, including exploring the origins of the feelings of rejection and their role in the client's current psychological difficulties. This will also impact on the way in which the client experiences their client-therapist relationship. The questions that arise from this are similar to those experienced in offline counselling settings. In FtF encounters, the therapist needs to assess the role that s/he is perceived to play by the client and incorporate this into the therapeutic process. The absence of cues or an over-edited, manipulated presentation of self in email communication could thus endanger a counsellor's accurate perception of the client's perceptions and needs. The role of time-delay in email communication thus starts to present itself as a double-edged sword: Whilst it may prove advantageous to the client to be able to construct and manipulate their self-presentation, to the counsellor, the additional time afforded the client may actually prove to be a hindrance to ascertaining their true needs for a successful outcome of the counselling relationship.

Time delays should therefore be clearly articulated to the client when beginning the email counselling process. In line with this proposal, research has suggested that clients may feel reassured if they can anticipate when they will receive an email reply (Dunn, 2012). Counsellors could consider the use of autoreply when emails are received in order for clients to be fully informed of the process. Research has examined the relationship between time to respond to emails within a counselling service, and whether the service was used at a later time for follow up support. Findings have also suggested that those with longer wait times are less likely to use the service again (Caffery & Smith, 2006). Additionally, interviews with online counsellors have highlighted concerns about how an online exchange may not easily transfer to offline contexts. Counsellors have suggested that once a client has engaged with a therapist in an online setting, it may take time to re-establish rapport offline (Dunn, 2012). From these discussions on email therapy, one might be of the opinion that the email therapeutic process is too fraught with difficulties to be considered useful. It might be the case that the therapist has to weigh up the benefits/costs of this mode of communication for each individual client. It might also be the case that the therapist then recommends a different form of online counselling to the client, regardless of monetary gain, such as that of online chat therapy (or instant messaging) online.

8.4   Online chat therapy (instant messaging)

Whilst online chat has similar characteristics and advantages to email therapy, a notable difference is that online chat or instant messaging operates as a synchronous form of communication. That is, client and therapist engage with each other in real time (Fenichel et al., 2002). Unlike with emails, if a client and therapist engage in therapy via instant messaging, they are required to book an appointment/time-slot and attend at that time (Goss & Anthony, 2003). However, similarly to email therapy, individuals remain able to hide their physical appearance from their therapist, and focus on the written exchange instead (Manhal-Baugus, 2001). Some individuals thus report a decrease in self-consciousness in instant messaging therapy (Haberstroh et al., 2007). Throughout this chapter we use the abbreviation IM to refer to both instant messaging and online chat.

Findings around the success of IM therapy are varied, with some research suggesting that IM holds a similar level of efficacy to FtF methods (Murphy et al., 2009). However, when outliers were removed from the study reported by Murphy et al., FtF counselling was shown to be superior in several areas. It was suggested that the individuals who formed the outliers in the study had less of an understanding of what to expect from the IM medium. It is therefore important to ensure that clients who use any form of online communication for counselling are adequately prepared for and understand the process. Research has also compared outcomes of online counselling to phone counselling. When 100 young people received a single session of phone counselling and 86 people received IM counselling, both demonstrated good outcomes for reducing psychological distress. However, phone counselling offered a significantly larger reduction in psychological distress (King et al., 2006). It was suggested that phone consultations are not necessarily superior in terms of emotional connection between client and therapist, but that differences in findings may be due to the smaller number of exchanges present within the chat medium (King et al., 2006). This is consistent with recent research which has suggested that if clients attend two IM sessions, more stages of counselling are progressed through in greater depth than in a single session, along with greater decreases in psychological distress (Dowling & Rickwood 2015). However, interviews with online counsellors have revealed that clients are more likely to cut to the chase within online chat than FtF sessions, due to the absence of immediate proximity of the counsellor and a perception of psychological safety (Bambling et al., 2008). This represents *benign disinhibition* (Suler, 2004), whereby individuals disclose more information at a more rapid pace online than offline. Whilst this may be due to *solipsistic Introjection* (Suler, 2004), it may also relate to a *minimisation of status and authority* of the counsellors (Suler, 2004). That said, perceptions of anonymity and decreases of perceived status of counsellors can lead to increases in aggressive behaviours of clients in IM (Dowling & Rickwood, 2014). Perceptions of minimisation of status and authority in online settings can also lead to *toxic disinhibition*, an increase in unpleasant behaviour which may not be displayed offline (Suler, 2004). It is therefore important that counsellors are trained in how best to respond to these processes within the online therapy environment (Dowling & Rickwood 2014). Understanding how these different psychological factors impact the online counselling relationship thus becomes pivotal to ensuring the psychological safety and understanding of both counsellors and clients online.

The decrease in inhibition and increase in perceived control may mean that clients in online chat settings can withhold information which may be needed by therapists, such as emergency contact details (Bambling et al., 2008). This is enabled by the increased ability to edit and selectively self-present (Walther, 1996), compared to FtF counselling. In ways similar to those outlined for email therapy, it is noted that a client's ability to edit responses is a form of client non-disclosure and is not always an advantage to the therapeutic process. For example, the client may move towards a higher level of denial than might ordinarily be the case in FtF counselling, and may not allow the therapist to have adequate information to consider the client's true depth of need. In addition, if clients choose to think too carefully about their input to the interaction, this *filtering* may lead to the spontaneous nature of the interaction becoming lost (Suler, 2000). Thus, whilst IM may be somewhat more spontaneous than email therapy, there remains the possibility of the client presenting to the therapist a distorted view of the self.

Research has also suggested that in IM therapists do not always use all the available counselling interventions available to them, and which they use offline (Mallen et al., 2011). Therapists have stated, for instance, that within online chat they walk a fine line between obtaining enough information and keeping clients engaged (Dowling & Rickwood, 2014). Clients have also reportedly found it easier to withdraw from a session online than in FtF settings (Fletcher-Tomenius & Vossler, 2009). In some cases, however, such instant withdrawal may be part of the social difficulty faced by the client, with the immediacy of withdrawal offering the counsellor little immediate opportunity to explore this issue with the client. This may offer an explanation for why some interventions (such as types of questions) are used more frequently than others within online settings. Research has noted that some interventions within online chat are used more than others because of concerns about how the client will respond, and that some types of interventions are therefore used to make up for the lack of non-verbal cues in IM (Mallen et al., 2011). Minimising the occurrences of misunderstandings is a real concern amongst therapists, and use of correct spelling and clear meaning is an important consideration within IM (Chester & Glass, 2006; Tate & Zabinski, 2004).

It might be expected that due to time pressures and for ease of communication, clients would use abbreviated or text speak in their online counselling sessions, and this might increase the potential for ambiguities. However, there appears to be a dearth of research examining how communication via text speak can impact on the ability to form a useful working alliance, and how it might impact adversely on ability to communicate, with a potential for misunderstandings. Haxell (2015) has, however, considered the use of a text-message-based support service for young people and noted that when using text-based communication, many issues are often relayed in a single message. This can lead to counsellors not having sufficient cues to guide their responses, such as the age of the client. Whilst counsellors may find the lack of cues in online chat difficult to work with, clients have also noted that the same phrases can be interpreted differently depending on the method of communication. For example, the seemingly innocuous phrase *"tell me more about that,"* may be perceived as patronising by the online client, who acknowledges that FtF, the question would have been interpreted differently (Haberstroh et al., 2007). Misunderstandings may thus occur because the spontaneous clarification individuals can offer in FtF interaction requires effort and skill to input to a text-only medium (Rochlen et al., 2004). Some clients, such as those who are sensitive to distortion of information and those who may become more distressed via text communications, may not be best-suited to this mode of communication for therapy (Yager, 2003). This type of intervention is therefore considered inappropriate for clients experiencing marked mental health difficulties (Rochlen et al., 2004), and suicidal or psychotic clients (Chester & Glass, 2006).

Whilst there may be problems relating to the lack of verbal cues for communication within text-based counselling methods, there is a suggestion that there are ways to compensate for the lack of non-verbal cues (Murphy et al., 1998). For example, practitioners have explained how they use acronyms and abbreviations as a means of conveying emotions, as well as emotional bracketing (Fletcher-Tomenius & Vossler, 2009). *Emotional bracketing* refers to the use of brackets around emotional material (Murphy et al., 1998). Therapists can thus employ emoticons along with the bracketing of emotional content to ensure the proper/accurate interpretation is portrayed to clients (Collie et al., 2000).

A further limitation of achieving a satisfactory therapeutic relationship may be the difficulty of overcoming any emotional vocabulary limitation with a client when using IM. Such limitations are likely to affect the expression of feelings in the online encounter. In addition, as is the case with email therapy, cues of body language and facial expression, as well as auditory and voice intonation cues that are prevalent in a FtF environment, will be absent in online chat exchanges. New ways of interpreting emotions and eliciting these from the client will therefore need to be part of a traditional counsellor's online skill set. Subtle cues will need to be looked for within any written narrative, email, or instant messaging. Direct questions may need to be asked. This will be particularly so if the counsellor experiences high levels of defensiveness in the online exchange. The counsellor may need to learn to elaborate their own expression of emotion in the online exchange to facilitate the client's expression of emotion (Anthony & Nagel, 2010). Alternatively, the counsellor may feel the need to switch to a mode of communication that might reinstate some of these cues, such as video communication.

8.5   Video-counselling

Video-counselling has received comparatively little research focus as against other forms of online counselling (Goss & Anthony, 2003) and yet it may be very valuable in eliminating difficulties around honesty and trustworthiness on both the client and the therapist side of the online therapeutic relationship. These are interactions which occur in real time, synchronously, between a client and therapist who can see and hear each other (Down, 2009). Unlike email and IM methods of therapy, cues to interaction remain visible, tone of voice, body language, and physical appearance of client and therapist all remain present, and ambiguities may be reduced (Simpson 2009; Suler, 2000). In some instances though, video counselling may provide clients with too many social cues (Suler, 2000) and their ability both to reflect and to engage effectively may become hampered by concerns, such as social responses of the therapist or maintaining eye contact. Indeed, it has been suggested that the online cues are different to those in offline FtF counselling, with a true depth of person-focused communication where subtle non-verbal cues can be seen, accurately interpreted, and responded to as in traditional FtF therapy missing in video-counselling (Quarto, 2011). One factor that may, however, prevail, and cause the client a lot of anxiety and concern is that they may want to come across as being likeable on screen. We know from research that individuals may be perceived as less likeable in video-mediated communications compared to FtF, and this may be related to a reduction in visual signals including eye gaze. It has been observed for instance that if individuals engaging in video communications gaze at the monitor, which will display the image of the other, the perception received by the communication partner is that of gazing downwards, this may give the perception of gaze avoidance, with potential for negative effect on perceived likeableness (Fullwood, 2007). This could affect the ways in which clients experience this type of counselling session. They may not experience counselling via videolink as positively as a FtF counselling session. Similar research has noted that participants recall more information when a conversational partner is looking directly at the camera than if gaze is perceived as being averted (Fullwood & Doherty-Sneddon, 2006). If individuals do not perceive their conversational partner as engaged, this may also impact on the therapeutic process.

One of the disadvantages of synchronous communications such as video-counselling is the potential for signal loss and link disruption. The therapeutic interaction could easily be impacted by loss of communication, which may make some clients feel as if they have been abandoned or rejected (Levy & Orlans, 2014). This would be particularly so with those who have insecure attachments and would undermine their sense of being listened to and nurtured in the online counselling environment. It is thus essential that from the very outset of the counselling relationship that plans for impaired or lost communication are drawn up between therapist and client. There also exists a technical disadvantage within video-mediated counselling: if the quality of the technology is poor due to slow refresh times and pixellated movement, low quality sound or distorted visual images can result in cues from body language being missed and the occurrence of sound delays (Lewis et al., 2003). These delays may impact negatively on the types of disclosures a client makes, leading to a poorer quality interaction (Down, 2009). This may then limit the therapeutic relationship and counselling progress, as the concern with technical issues overrides the therapeutic encounter. Nonetheless, recent research has suggested that counselling conducted via video link can be as beneficial to clients as FtF counselling. Stubbings et al. (2013) compared outcomes for 30 participants who engaged in cognitive behavioural therapy (CBT) in person or by video link. Both processes were associated with decreases in depression, anxiety, and stress, with ratings of working alliance and satisfaction ratings comparable between the two conditions. Similarly, it has been suggested that video-conducted therapy did not differ from FtF in terms of client ratings of helpfulness. Clients were also more likely to miss FtF sessions than video link sessions (Veder et al., 2014).

Webcam communications may be more suitable for individuals who do not find benefit from a text-driven form of therapy, such as those who find typing difficult or who are less comfortable expressing themselves outside of direct verbal encounters (Suler, 2000). Individuals may prefer this method since it offers a more immediate method of communication than the asynchronous methods of email and IM (Fenichel et al., 2002) and may create an increased feeling of presence with the therapist compared to other modalities (Skinner & Zack, 2004). Clients who need a greater sense of perceived personal connection by being able to see the counsellor and express themselves through physical gesture and tone of voice may prefer this mode of communication, although further research into this area is required.

Thus far we have focused on the mode of communication and how therapy might be delivered online. We have also considered some of the advantages and disadvantages of the three different types of connection outlined (email, IM, Video). Our attention in the final section of this chapter turns briefly to considering some of the different types of therapy that might be carried out across these different modes of communication.

8.6   Types of online counselling

A common definition of counselling and psychotherapy might include the use of talking therapies delivered by trained professionals to clients. The client-therapist interaction might be a short term experience, but could equally last for years, with this traditional process generally revolving around the professionally trained therapist and a single client (Post et al., 2002). For a more thorough outline of the ways in which therapist and client come together to engage in what has classically been referred to as offline talking therapies, the reader is directed to the British Association for Counselling and Psychotherapy (BACP) website (http://www.bacp.co.uk). The combined effort of client and therapist incorporates a broad range of approaches from varying psychological traditions (Beck, 2011). Explaining the tenets of the wide and varied therapies available is beyond the scope of this chapter. Suffice it to say that some traditional counselling approaches include *cognitive behavioural, psychodynamic,* and *humanistic* therapies (Woolfe et al., 2009).

8.6.1   Cognitive behavioural therapy and the person-centred approach

The central focus of *cognitive behavioural therapy* (CBT) is, for instance, to understand an individual's perception and interpretation of events in relation to overt behaviours and actions (Beck, 2011) and has been considered to offer an ideal approach for online therapy (Andersson,

2009; Hughes, 2008). CBT uses the activity of homework which may be given to the client in written form for them to carry out after the session and to report back on for the next session. Within online contexts this homework could be emailed before the next session to alert the online counsellor of the progress of the therapy and allow suggestions of further practical homework to be carried out to build on the previous session. The client might feel more supported by counsellor prompts via email to continue with the homework. Some CBT-influenced therapists use psycho-educational websites to supplement such individual therapy sessions, which may also allow modelling. In an online environment, this can be achieved through a videoed presentation or in a podcast (Andersson, 2009).

An example of a *psycho-educational approach* is the work of Hughes (2008), who carried out a CBT-influenced online intervention for bulimia. Here, the clients were provided with website information that would ordinarily be given in a work book context. The information was read online, with CBT approaches supplemented through weekly group chat sessions with a therapist. Comparing this approach to other online approaches, Anthony and Nagel (2010) emphasised the importance of clients carrying out practical work.

Anthony and Nagel (2010) also notably support application of the *person-centred approach* in online counselling, given that *"its intrinsic belief that it is the client and therapist relationship that is central to the work ... and the belief in the work being client-led rather than practitioner led ..."* (p. 13). Bearing in mind that the central tenets of person-centred approach, known as the core conditions, are generally seen as being a foundation for the success of any good counselling approach, it might be the case that a combination of CBT and the person-centred approach work best online. To elaborate, Evans (2009) notes that particular skills are required in trying to establish the core conditions of *empathy, congruence, unconditional positive regard*, and, importantly, further conditions related to counsellor and client being in psychological contact (Rogers, 2005). The latter may or may not require some counsellors to be in FtF contact. For others, particularly those who have confidence in online counselling approaches, such psychological contact may be achieved through the implied empathic link as expressed through the written word and/or webcam exchanges. As Evans (2009) argues *"Empathy is a key skill in both online and face-to-face supportive interventions ..."* (p. 55) thus suggesting an importance of empathy regardless of the mode of communication and type of counselling taking place.

Let us consider in more detail the role of empathy when using these approaches in a text-based mode of online communication. One risk that might ensue is that of counsellors simply cutting and pasting elements of a client's communication as a way of mechanically reflecting back the client's perspectives (e.g., Pelling & Reynard, 2000). Whilst the counsellor might perceive this to be a useful tool for reiterating and interpreting the content, this could infuriate and/or frustrate the client, who may feel that the counsellor is simply providing repetition rather than advice or understanding. Anthony and Nagel's (2010) counter-argument to this notion would suggest that in a positive text-based relationship, it is the technology that is the primary feature of the counselling process, whilst the actual communication takes on a secondary role. Also, it is worth noting that using a person-centred approach to counselling offers the key benefit of the counsellor having the client's actual written communication to fall back on rather than needing to paraphrase, thus ensuring a higher level of accuracy in the therapist's response to the client. An example of a text-based exchange where empathy may be conveyed is demonstrated in the following response to a 35-year-old client, who feels trapped and angry due to experiencing physical disability related to a road traffic accident at the age of 25, and experiences feelings of rage related to loss of occupation, income, and their previous extensive social connections:

> You have experienced significant and unpredicted changes in your life, and nothing could have prepared you for this. You feel a huge amount of anger, resentment, and distress due to the level of loss that you have experienced. It is difficult to cope with the enormity of the change that you are experiencing. You do not know where to start. It feels overwhelming. You tell me that you feel let down and abandoned by friends who were once close, who now no longer take the time to call. You describe yourself as losing contact with your long-term peer group and questioning why you should carry on with such hurt.

The text response from the online therapist illustrates the manner in which they have been able to capture and reflect the client's position using text-based skills to convey empathy. Empathy would thus need to be maintained, and any repetition, re-posting or reiteration of content be very carefully considered by the therapist to meet the empathy requirements of the client. This demonstrates more of an advantage of text-based over video-based approaches to delivering this type of therapy. In text-based approaches the words exist in a physical presence and can either remain on screen or be printed for the client to review when feelings of anger and loss again surface, to enable the client to feel a reconnection with the therapist and to experience acceptance and support which they perceive themselves not to have elsewhere.

8.7   Which therapy, when, where, and how?

Regardless of how therapy is conducted and the medium employed, it is dependent upon forming a therapeutic working relationship between the client and therapist (Jones & Stokes, 2008). Online counselling, as with any counselling modality, has to be best suited to the clients' individual needs. It is therefore rather simplistic to suppose that we might be able to advocate one form of counselling being best suited to a given mode of communication and another form to an alternative mode of communication. That said, it has been suggested that some cognitive therapy approaches may adapt more easily to an online environment due to their direct questioning approach, rather than those therapies which use a more reflective approach such as client centred counselling or psychodynamic approaches (Jones & Stokes, 2008), possibly partially due to difficulties in rephrasing questions online. In opposition to Jones and Stokes (2008), Anthony and Nagel (2010) argue that regardless of whether therapy is delivered online or FtF, the therapeutic process remains the same. If this is the case, then there should be no best match between type of therapy and mode of delivery. The success and satisfaction of therapy for the client can only come about if the therapist is able to establish a mutual relationship connection involving trust and respect to develop the relationship with the client to begin to explore the issues that they bring into therapy. This should then take precedence when deciding on which type of therapy to provide and how to provide it. With the emphasis on the nature of the client-counsellor relationship, Hick and Bien (2008; see also Jones & Stokes, 2008; Paul & Haugh, 2008) have argued that it is the therapeutic relationship between the client and the practitioner that is the primary issue in enabling a satisfactory counselling outcome, rather than the impact of a particular counselling approach. From our very brief consideration of the different types of therapy in relation to different modes of online communication, it would appear that there remains a considerable amount of work to be carried out in order to establish when certain types of therapy work best, and for which types of client, via diverse modes of communication.

8.8   Theoretical considerations

Online counselling is a very real and relevant phenomenon, regardless of how it is delivered and which type of counselling is considered. There are a number of cyberpsychological theories that might lend themselves to further exploring some of the questions raised by this chapter. The *hyperpersonal model of computer-mediated communication* (HPCMC) (Walther, 1996) may assist in the understanding of the benefits clients perceive in an asynchronous text-based method, where physical appearance does not impede interactions (Goss & Anthony, 2003). Contrary to popular models of its time, the HMCMC model suggested that online communication can be advantageous in many situations because of its removal of situational and dispositional cues. Therefore, clients might feel less inhibited in online communications because they can promote a more ideal version of self to the therapist. They can selectively choose which information they might want to share, especially, as already mentioned, when they are creating, editing, and re-editing a version of self to be presented via asynchronous modes of communication, such as an email to a therapist. The HPCMC model also suggests that people will seek out further information on a communications partner if they feel that they perceive a similarity to that person, and if they anticipate a future interaction with that person. In the counselling sense, it might be suggested that real or imagined perceptions of similarity of the client to the therapist might erroneously evoke a sense of kinship in the client that could prove detrimental to the therapeutic relationship. According to HPCMC then, it might be the case that clear therapy boundaries need to be established prior to entering a therapeutic relationship online. In continued development of the HPCMC model, Walther (2007) recently considered other factors that impact upon online interactions. A key feature of these considerations was that of *individual motivation*. This is linked to anticipation of taking the counselling relationship offline, for example. If a client does not anticipate a future physical interaction with the therapist, then they might convey a less considered, more accurate version of their self to the therapist. If they do anticipate an offline interaction, they might work harder to construct messages that paint an optimised picture of the self to the therapist. The potential transition of a therapeutic relationship to offline counselling might therefore also cause concern, especially if the client engages in enhanced manipulation of self-presentation without anticipation of an offline interaction. Clearly, these considerations raise questions as to the authenticity of clients in online counselling relationships. Indeed, it might even be the case that the *counsellor* misrepresents themselves online if they have absolutely no intention of meeting with a client in future offline. In such instances, charlatan counsellors might offer services based on stock replies in the most asynchronous and impersonal ways possible to ensure nothing other than a monetary gain from the process. Nonetheless, for the purposes of this chapter we are assuming that the therapeutic relationships under scrutiny are of a genuine nature.

Another way in which we might apply this theory to understanding online counselling is to consider the synchronicity of online counselling. If engaging in asynchronous communication, for instance, a client will have much more control over their self presentation, as they will have the time to create and edit their communication. For the counsellor, this might present a frustrating scenario in which s/he struggles to ascertain the true characteristics of the client. The counsellor might experience increased difficulty in getting to the heart of the client's problems if s/he is constantly presenting an optimal version of the self.

Since the creation of the HPCMC model, video communication (VC) has rapidly become a more popular mode of online communication. It could be suggested that VC reinstates many of the cues that are absent in text-based online communications. The model may thus struggle to explain the advantages and disadvantages that might prevail in VC counselling sessions. Whilst some research has suggested that due to the potential for perceptions of gaze aversion, VC might not fully alleviate problems with reduced cues that might ensue in text-based communications (Fullwood, 2007), a similar level of therapeutic efficacy between the different modes of communication has been reported (Veder et al., 2014). Further research is clearly necessary to gauge whether this model can be used to explain how when and where the use of online counselling might be more or less advantageous to different clients.

8.9   Concluding comments

There are many forms of online therapy, only a few of which have been touched upon in this chapter. One of their shared advantages is that their delivery enables access to a wide variety of people, who might otherwise be geographically limited in their access to traditional counselling (Sussman, 2004), or personal mobility limitations (Maples & Han, 2008) and/or counselling process preferences. As such, online therapies might be a useful form of intervention for those who are anxious, reluctant, or unable to seek help in a FtF setting (Lange et al., 2003). It is interesting to note that recent research has suggested that clients are increasingly using online counselling as an adjunct to FtF counselling rather than as an alternative. A number of factors have, however, come to light in this chapter that need consideration for the successful outcome of online therapeutic processes. These include, but are not limited to, individual client's needs and their personal preferences for synchronous or asynchronous communications as well as their preference for cue-rich (e.g., VC) or cue-absent (e.g., text-based) communications that enable a more or less manipulated presentation of the self to a therapist. It would appear that online counselling might be useful in many situations, but this should not be taken to suggest the future eradication of FtF therapeutic interventions. There remains a very real place for traditional counselling methods. It might, thus, be counterproductive to think that one method should be preferred over the other. Rather, we should think of this as an emerging area of online behaviour that has the scope and potential to be sufficiently influenced by the rapidly growing area of cyberpsychology to provide a service beneficial to all involved.

8.10   References

Andersson, G. (2009). Using the internet to provide Cognitive Behavioural Therapy. *Behaviour, Research and Therapy, 47*(3), 175–180.

Anthony, K., & Nagel, D. (2010). *Therapy online: A practical guide.* London: Sage.

Bambling, M., King, R., Reid, W., & Wegner, K. (2008). Online counselling: The experience of counsellors providing synchronous single-session counselling to young people. *Counselling and Psychotherapy Research, 8*(2), 110–116.

Beck, J. (2011). *Cognitive behavior therapy, 2nd edition: Basics and beyond.* New York, London: Guilford Press.

Bor, R., Gill, S., Miller, R., & Parrott, C. (2004). *Doing therapy briefly.* Basingstoke and New York: Palgrave Macmillan.

Caffery, L., & Smith, A. (2006). The relation between response time and the re-utilization of an email based counselling system. *Journal of Telemedicine and Telecare, 12*(3), 20–22.

Chechele, P., & Stofle, G. (2003). Individual therapy online via email and internet relay chat. In Goss, S. & Anthony, K. (Eds.), *Technology in counselling and psychotherapy: A practitioner's guide.* Houndmills, Basingstoke, Hampshire: Palgrave Macmillan.

Chester, A., & Glass, C. (2006). Online counselling: A descriptive analysis of therapy services on the internet. *British Journal of Guidance & Counselling, 34*(2), 145–160.

Collie, K., Mitchell, D., & Murphy, L. (2000). Skills for online counseling: Maximum impact at minimum bandwidth. In Bloom, J. & Walz, G. (Eds.), *Cybercounseling and cyberlearning: Strategies and resources for the millennium.* Alexandria, VA: American Counseling Association.

Dowling, M., & Rickwood, D. (2014). Experiences of counsellors providing online chat counselling to young people. *Australian Journal of Guidance and Counselling, 24*(2), 183–196.

Dowling, M., & Rickwood, D. (2015). Investigating individual online synchronous chat counselling processes and treatment outcomes for young people. *Advances in Mental Health, 12*(3), 216–224.

Down, P. (2009). *Introduction to videoconferencing.* Retrieved August 25, 2015 from https://www.ja.net/sites/default/files/Introduction%20to%20Videoconferencing.pdf

Dunn, K. (2012). A qualitative investigation into the online counselling relationship: To meet or not to meet, that is the question. *Counselling and Psychotherapy Research,12*(4), 316–326.

Evans, J. (2009). *Online counselling and guidance skills: A practical resource for trainees and practitioners.* London: Sage.

Fenichel, M., Barak, A., Zelvin, E., Jones, G., Munro, K., Meunier, V., & Walker-Schmucker, W. (2002). Myths and realities of online clinical work. *CyberPsychology & Behavior, 5*(5), 481–497.

Fletcher-Tomenius, L., & Vossler, A. (2009). Trust in online therapeutic relationships: The therapist's experience. *Counselling Psychology Review, 24*(2), 24–34.

Fullwood, C. (2007). The effect of mediation on impression formation: A comparison of face-to-face and video-mediated conditions. *Applied Ergonomics, 38*(3), 267–273.

Fullwood, C., & Doherty-Sneddon, G. (2006). Effect of gazing at the camera during a video link on recall. *Applied Ergonomics, 37*(2), 167–175.

Goss, S., & Anthony, K. (2003). *Technology in counselling and psychotherapy. A Practitioners' Guide.* Houndmills, Basingstoke, Hampshire: Palgrave Macmillan

Haberstroh, S., Duffey, T., Evans, M., & Trepal, H. (2007). The experience of online counselling. *Journal of Mental Health Counseling, 29*(3), 269–282.

Haxell, A. (2015). On becoming textually active at Youthline, New Zealand. *British Journal of Guidance & Counselling, 43*(1), 144–155.

Hick, S., & Bien, T. (2008). *Mindfulness and the therapeutic relationship.* New York: Guilford Press.

Hughes, T. (2008). Internet based study of Cognitive Behavioural Therapy for Bulimia to be conducted by UNC, WPIC. *Medical News Today.* Available at www.medicalnewstoday.com/articles/121638.php

Jones, G., & Stokes, A. (2008). *Online counselling. A handbook for practitioners.* Palgrave Houndmills, Basingstoke, Hampshire: Macmillan.

King, R., Bambling, M., Reid, W., & Thomas, I. (2006). Telephone and online counselling for young people: A naturalistic comparison of session outcome, session impact and therapeutic alliance. *Counselling and Psychotherapy Research, 6*(3), 175–181.

Lange, A., Rietdijk, D., Hudcovicova, M., Van de Ven, J. P., Schrieken, S., & Emmelkamp, P. (2003). Interapy. A controlled randomized trial of the standardized treatment of posttraumatic stress through the Internet. *Journal of Consulting and Clinical Psychology, 71*(5), 901–909.

Levy, T., & Orlans, M. (2014). *Attachment, trauma and healing: Understanding and treating attachment disorder in children, families and adults* (2nd ed.). London and Philadelphia: Jessica Kingsley Publishers.

Lewis, J., Coursol, D., & Wahl, K. (2003). *Researching the cybercounselling process: A study of the client and counselor experience.* Retrieved June 15, 2015 from http://files.eric.ed.gov/fulltext/ED481145.pdf

Mallen, M., Jenkins, I., Vogel, D., & Day, S. (2011). Online counselling: An initial examination of the process in a synchronous chat environment. *Counselling and Psychotherapy Research, 11*(3), 220–227.

Manhal-Baugus, M. (2001). E-therapy: Practical, ethical, and legal issues. *Cyberpsychology & Behavior, 4*(5), 551–563.

Maples, M., & Han, S. (2008). Cybercounseling in the United States and South Korea: Implications for counseling college students of the millennial generation and the networked generation. *Journal of Counseling & Development, 86*(2), 178–183.

Murphy, L., & Mitchell, D. (1998). When writing helps to heal: E-mail as therapy. *British Journal of Guidance & Counselling, 26*(1), 21–32.

Murphy, L., Parnass, P., Mitchell, D., Hallett, R., Cayley, P., & Seagram, S. (2009). Client Satisfaction and outcome comparisons of online and face-to-face counselling methods. *British Journal of Social Work, 39*(4), 627–640.

Paul, S., & Haugh, S. (2008). The relationship, not the therapy? What the research tells us. In. S. Haugh & S. Paul (Eds.), *The therapeutic relationship: Perspectives and themes*. Ross on Wye: PCCS Books.

Pelling, N., & Reynard, D. (2000). Counselling via the internet, can it be done well? *The Psychotherapy Review, 2*(2), 68–72.

Pollock, S. (2006). Internet counseling and its feasibility for marriage and family counseling. *The Family Journal: Counseling and Therapy for Couples and Families, 14*(1), 65–70.

Post, A., Borgen, W., Amundson, N., & Washburn, C. (2002). *Handbook on career counselling: A practical manual for developing, implementing and assessing career counselling services in higher education settings*. Available at http://unesdoc.unesco.org/images/0012/001257/125740e.pdf

Proudfoot, J. (2004). Computer-based treatment for anxiety and depression: Is it feasible? Is it effective? *Neuroscience & Biobehavioral Reviews, 28*(3), 353–363.

Quarto, C. (2011). Influencing college students' perceptions of videocounseling. *Journal of College Student Psychotherapy, 25*(4), 311–325.

Rochlen, A., Zack, J., & Speyer, C. (2004). Online therapy: Review of relevant definitions, debates, and current empirical support. *Journal of Clinical Psychology, 60*(3), 269–283.

Rogers, C. (2005). *The Carl Rodgers Reader* edited by Kirshenbaum and Henderson. London: Constable & Robertson.

Simpson, S. (2009). Psychotherapy via videoconferencing: A review. *British Journal of Guidance & Counselling, 37*(3), 271–286.

Skinner, A., & Zack, J. (2004). Counseling and the Internet. *The American Behavioral Scientist, 48*(4), 343–446.

Stubbings, D., Rees, C., Roberts, L., & Kane, R. (2013). Comparing in-person to videoconference-based cognitive behavioral therapy for mood and anxiety disorders: Randomized controlled trial. *Journal of Medical Internet Research, 15*(11), e258.

Suler, J. (2000). Psychotherapy in cyberspace: A 5-dimensional model of online and computer-mediated psychotherapy. *Cyber Psychology & Behavior, 3*(2), 151–159.

Suler, J. (2004). The online disinhibition effect. *Cyberpsychology and Behavior, 7*(3), 321–326.

Sussman, R. (2004). Counseling over the Internet: Benefits and challenges in the use of new technologies. In Waltz, G. & Kirkman, C. (Eds.), *Cyberbytes: Highlighting compelling uses of technology in counseling* (pp. 17–20). Retrieved August 18, 2015 from http://files.eric.ed.gov/fulltext/ED478216.pdf

Tate, D., & Zabinski, M. (2004). Computer and Internet applications for psychological treatment: Update for clinicians. *Journal of Clinical Psychology: In Session, 60*(2), 209–220.

Veder, B., Pope, S., Mani, M., Beaudoin, K., & Ritchie, J. (2014). Employee and family assistance video counseling program: A post launch retrospective comparison with in-person counseling outcomes. *Medicine 2.0. 3*(1), e3.

Walther, J. (1996). Computer-mediated communication: Impersonal, interpersonal, and hyperpersonal interaction. *Communication Research, 23*(1), 3–43.

Walther, J. (2007). Selective self-presentation in computer-mediated communication: Hyperpersonal dimensions of technology, language, and cognition. *Computers in Human Behavior, 23*(5), 2538–2557.

Wellman, B., & Haythornthwaite, C. (2002). *The internet in everyday life.* Oxford: Blackwell.

Woolfe, R., Strawbridge, S., Douglas, B., & Dryden, W. (2009). *Handbook of counselling psychology.* London: Sage.

Wright J., & Bolton G. (2012) *Reflective writing in counselling and psychotherapy.* London: Sage.

Yager, J. (2003). Monitoring patients with eating disorders by using e-mail as an adjunct to clinical activities. *Psychiatric Services, 54*(12), 1586–1588.

9
Romantic Relationships and Online Dating

Nicola Fox Hamilton

9.1 Introduction

The influence of technology in our lives has seeped into nearly every aspect of how we relate to others. We connect with our friends and family through text, email, social networking sites (SNS), and instant messaging to name but a few. Through a variety of online platforms we seek old and new friends, business partnerships and collaborations, employers and employees and of course, we seek candidates for those relationships most dear to us, romantic relationships. This chapter cannot attempt to address the vast area of how technology changes the ways in which we interact in all of our relationships, but rather will focus on the influence of technology and the Internet on our romantic relationships, in particular how we find those relationships through online dating.

Meeting online has displaced other ways to meet, and has become the third most likely way for people to connect with a new partner, with one in five relationships now starting online (Rosenfeld & Thomas, 2012). As such, it is important to understand how people present themselves and perceive others, how we relate to each other during the process of online dating, and how we view the positive and negative aspects of the overall experience. It is also important to understand that not everyone experiences online dating in the same way, and the experiences of different groups, such as men, women, and LGBT daters will be examined. There are significant quantities of research being undertaken on the subject and the focus of this chapter will be to examine and review the academic literature relating to the psychology of online and app-based dating to determine what insights have been gleaned for both daters and businesses involved in online dating in over a decade of studies. The chapter will address the process of online dating from a dater's perspective, for example, looking at the effects of photographs and language on attraction online, what we find attractive and how we make partner choices, how authenticity is determined and trust developed, and the prevalence of deception and how daters attempt to counteract it. Recommendations will be made to help users improve their experience dating online. Online daters' experiences of dating sites and applications, while generally positive, also comprise a range of frustrations and difficulties that are imposed by the limitations of the platforms. The research conducted on online dating to date suggests a number of areas in which improvements and additions can be made to sites in order to help users overcome these difficulties. These include interactivity, social possibilities, and interface design.

9.2   Who dates online and where do they meet?

Online dating has grown exponentially across many areas of the world since first emerging around 1997, and meeting partners online exploded with the growth of Web 2.0 technologies. Of cohabiting couples across 18 countries, including Europe, Brazil, and Japan, nearly one third have experience of online dating, and of those who have begun relationships since 1997, 15% are in a relationship that began online (Hogan et al., 2011). Globally, there are differences in the adoption of online dating and other online spaces for finding romantic relationships, with Internet access and penetration playing a part. Northern European inhabitants are more likely than their southern counterparts to find love online. A third of Germans in relationships that have begun since 1997 found their partner online, whereas Greeks are the least likely of all to have found a partner online, with only 15.5% having done so. Online dating is used alongside and as a complement to more traditional means of meeting potential partners, such as introductions through friends, at work, or meeting in bars and other public spaces (Rosenfeld & Thomas, 2012). However, there has been an overall decline in these more traditional routes in general, and specifically in ways of meeting such as through church or community or hobby groups.

The social stigma that once existed around online dating has reduced considerably over time, but there is still a significant minority who consider finding a partner online to be an act of desperation (Smith & Duggan, 2013). Interestingly, attitudes towards online dating can be changed by experience of the technology and engendering trust and confidence in it. Those who use it at all, even without success, tend to rate the technology more positively. Additionally, those who are aware of people within their social circles who have used it are also more likely to accept it and to try it themselves (Hogan et al., 2011). This is illustrated by the finding that in small towns in rural areas, word of mouth and experiences of friends or family using online dating can have a significant impact on the number of online daters in an area, producing a cluster of interrelated success stories (Mascaro et al., 2012).

Dating sites are the most common online spaces in which people meet romantic partners, with up to half meeting in this way (Hall, 2014; Hogan, et al., 2011; Smith & Duggan, 2013). Hogan et al. (2011) found that different cultures have differing levels of dating site usage; for example, Germans are most likely to meet through a dating site when they meet online, where are Brazilians are most likely to meet though social networking sites (SNS). SNSs are growing quickly in popularity as a means to meet, particularly among some minority groups such as African Americans in the United States (Hall, 2014). There are particular reasons for using SNS for finding romance, including the credibility and depth of information in profiles anchored to offline selves and the feeling that it is a more natural setting where people can meet through interests in common rather than the deliberate construction of identity in online dating (Lee & Bruckman, 2007). Other common places to meet online include chatrooms, though the popularity of these has decreased with the rise of social networking sites, and online communities such as discussion groups, virtual worlds, and online multiplayer games are also used regularly (Hogan et al., 2011).

Geosocial-networking applications such as Tinder and Grindr, also known as People Nearby Applications (PNAs), are a recent addition to the myriad technologies that can assist us in meeting new people, and in particular romantic partners. They are designed to allow people to meet prospective partners based on their proximity and on how attractive they find them. Gay men were early adopters of these applications and many use them regularly (Van De Wiele & Tong, 2014). Goedel and Duncan (2015) found that gay men who use these applications spend a considerable amount of time, on average 1.31 hours per day, doing so. Tinder is a popular new PNA for both heterosexual and homosexual users, although there is a lack of data on Tinder user experiences due to its recent emergence.

9.3   What characteristics do online daters have?

Certain innate traits make it more likely that an individual will try online dating. People with higher rejection sensitivity – differing levels of the tendency to "anxiously expect, readily perceive, and overreact to rejection" (Downey & Feldman, 1996, p. 1338), and lower self-esteem – the confidence that a person has in their own worth, are more likely to be online daters (Aretz et al., 2010; Blackhart et al., 2014). This may be because the fear of rejection is lowered online (Couch & Liamputtong, 2008; Rosenfeld & Thomas, 2012). Fear of rejection is reduced in a number of ways, such as the removal of the need to ask for a date face-to-face, and daters have some confidence that the majority of users are single and looking for love. The stigma of being rejected is reduced by the anonymity of the medium, and negative or non-responses can be interpreted in more favourable ways by the dater to soften the blow of rejection. For example, someone could interpret a lack of response to a message in a number of ways; the dater has left the site, they have met someone else, or they are too busy with an influx of new messages to reply to everyone (Rosenfeld & Thomas, 2012). On the other hand, lower anxiety around dating can also lead to more use of dating sites (Valkenburg & Peter, 2007). This supports the rich-get-richer hypothesis: that those who are confident daters offline are also confident daters online, and they look at online dating as an additional path for meeting a partner, rather than as a method of compensating for deficiencies in their social skills.

9.4   What drives people to date online?

There are a variety of motivations driving use of online dating applications and websites. These reasons vary in intensity across the particular platforms that people or groups of daters use, but have many similar characteristics, including finding romantic or sexual partners, making friends or being social, alleviating boredom, and constructing identity (Clemens et al., 2015; Couch & Liamputtong, 2008; Goedel & Duncan, 2015; Van De Wiele & Tong, 2014).

One of the primary reasons that people use online dating sites and other online platforms to meet new romantic partners is that they are experiencing a lean dating market offline. By the time that people reach their late 30s, approximately 80% of their peers are in relationships, leading to a lack of eligible partners to choose from, particularly in suburban and rural areas. This is exacerbated further for single LGBT romance-seekers. They already comprise a small proportion of the population and again many are already in partnerships, leading to a deficit of choice (Rosenfeld, 2010). In practical terms, online dating is seen as convenient, easy, and accessible. It opens a wide pool of potential participants and can involve less effort and cost to the user (Henry-Waring & Barraket, 2008), although typically men still carry the weight of the responsibility for initiating relationships online, which can be effortful (Smith & Duggan, 2013). Clemens et al. (2015) found that homosexual users seek a significantly broader number of uses of online dating than heterosexual users, particularly when using PNAs. While finding relationships and sex partners are key motivations, distraction, entertainment, and friendship are also motivations for using dating sites. Women are less likely to use dating sites for casual sexual encounters, but use them more than other groups to be social; and alleviating loneliness is a key concern of older daters (Clemens et al., 2015).

The ability to screen for desired characteristics is considered a positive feature of dating sites, particularly when it is known that people prefer to date others who are similar to themselves. This is known as *homophily*, the desire to bond with others who have similar characteristics to us. Homophily has been demonstrated in online dating, looking at both desired characteristics, and actions in contacting people on dating sites (Fiore et al., 2010; Hitsch et al., 2010; Morgan et al., 2010). Attributes such as marital status and desire for children are important areas where daters like to have similar attitudes and values, but also physical build and attractiveness, ethnicity, level of education, and smoking habits are some of the criteria where similarity is desired. When daters describe themselves as having certain characteristics relevant to their lifestyle, physical appearance, or personality, they are also likely to specify those qualities as required in a romantic partner (Morgan et al., 2010). Hence seeking out similar others to approach on dating sites, and using information about similarity in initial messages to attract attention, should lead to more success.

9.5   Online dating and self-presentation

Self-presentation is the consideration of how you present yourself to an audience to create a desired impression. This is often, but not always, used to create a positive impression, but can also be used to create a less positive appearance, such as being dangerous or edgy (Fullwood, 2015). In online dating people typically strive to create an attractive profile presentation. However, daters must also balance the ultimate goal of meeting face-to-face offline with the wish to market themselves as desirable online (Ellison et al., 2006; Whitty, 2008). Dishonesty can be quite common on online dating sites, particularly around physical descriptions such as height and weight, and areas of personality such as agreeableness and emotional stability. However most daters do not want to deceive and believe themselves to be truthful in their self-presentation and rationalise lying when the lies are about achievable discrepancies (Guadagno et al., 2012; Ellison et al., 2012; Whitty, 2008). A range of possible selves are drawn upon in creating dating profiles. Current theory of self-identity suggests that rather than having a unitary idea of the self, we each have different self-constructs comprised of a number of rather distinct and varied self-identities. These can include the actual self (who we currently are), the ideal self (who we would like to be) and the ought self (who we feel we should be) (Higgins, 1987). In creating dating profiles, strongly identified-with positive aspects of past and future selves, and aspects of the current self are employed in profile creation (Ellison et al., 2012; Whitty, 2008). For example when being deceptive, daters may draw upon anticipated future selves in their presentation rather than presenting an absolutely inaccurate portrayal of their present self. It is considered more acceptable to lie about going to the gym regularly when the dater anticipates that they will do so in the near future, than lying about height which is unchangeable. Ellison et al. (2012) posit that the dating profile is a psychological contract; a promise that "future face-to-face interaction will take place with someone who does not differ fundamentally from the person represented by the profile" (p. 12).

Creating a very positive impression in a profile can have a variety of consequences. An overly positive profile, where other dater feels it represents an ideal self rather than the actual self, is viewed as less attractive than a more realistic profile (Norton et al., 2007). Match.com has attempted to overcome the inclination for excessively positive profiles with the inclusion of a "love your imperfections" section that encourages people to share their flaws and foibles. However research shows that while greater disclosure of realistic positive information in a dating profile is successful in attracting others, greater honesty is not (Gibbs et al., 2006). This may mean that a profile that only presents the positive aspects of a person and not any of their flaws can result in many suitors on a dating site, but they may not be as well matched as suitors who are also aware of the flaws of the dater and still find them attractive. Hence a positive, but realistic profile may be the best chance of finding a successful match (Fiore et al., 2008; Gibbs et al., 2006).

9.6 Photographs and attraction

Online dating profiles typically consist of a profile picture, with perhaps a small gallery of supporting images; an open ended piece of text describing the dater and possibly some characteristics of their preferred partner; as well as a number of fixed choice set questions about searchable characteristics of the dater such as height, body type, number of children and smoking status. There is a considerable body of research to support the fact that physical attractiveness is one of the most important characteristics in romantic partner choice both online and offline (Cloyd, 1977; Fiore et al., 2008). Thus, it is not surprising that the photograph carries the most weight in determining the attractiveness of the overall dating profile. The *halo effect* is common in online dating, where a positive impression of a photograph leads to a positive overall impression of the profile and vice versa (Fiore et al., 2008). Almost everyone values physical attractiveness, most of all attractive people (Fiore et al., 2008). Interestingly, photographs rated highly attractive are considered to be less authentic by daters, but this does not stop daters from being more deceptive in their self-presentation in response to a highly attractive individual in order to increase their chance of a match (Lo et al., 2013). Photographs play an important additional role in the dating profile by validating claims made by the dater. For example, a photograph showing the dater engaged in bungee jumping provides evidence to support a claim of adventurousness in the profile. The rise of PNAs has shifted the focus even more towards photographs, as profiles tend to be pared back and contain considerably less information than traditional dating sites. Dating sites often offer guidelines on choosing photographs, and some, like Plenty of Fish, require that at least the profile picture show the dater's face. There is little peer-reviewed research on the most effective photographs, but as they are the primary method of determining attractiveness, the common tips provided by dating sites make sense. Some common recommendations include: using high quality photographs with natural light rather than flash photography, making the dater the central focus of the picture with their face clearly visible and including photographs that validate information mentioned in the profile, such as taking part in favourite activities.

9.7   Computer-mediated communication

Computer-mediated communication (CMC) imposes its own constraints on self-presentation in the dating process. Although photographs are a crucial component of the profile, the written aspects are also key to attracting suitors. The lack of social cues in CMC, such as body language, tone of voice and facial expressions, along with the *asynchronous* (delayed response) nature of it can lead to *hyperpersonal communication* (Walther, 1996). The lack of cues mean that impressions of the other person must come from what they write and how they write it. The sender of a message has time to consider and edit the message that they are sending, carefully deciding how to present themselves in an optimal fashion. Both parties have time to reflect and prepare their messages before they reply to each other. This can result in an intensified feedback loop where selectively composed and perceived messages are understood as positive feedback and can lead to heightened intimacy and positive impression formation (Walther, 1996). When this occurs the levels of emotion can be as high or higher than face-to-face communication (Walther, 1996). In conjunction with this, the online disinhibition effect can lower social inhibitions online and allow people to speak and act more freely online than they would in a face-to-face situation (Suler, 2004). Many online daters feel more able to express themselves freely online than face-to-face and disclose more personal or emotional information than they might otherwise, again intensifying the emotional interaction between communication partners (Henry-Waring & Barraket, 2008). While this can lead daters to believe that they have formed a connection with their communication partner, extended online communication can in fact lead to less successful outcomes offline than meeting face-to-face more promptly. Ramirez et al. (2015) found that between 17 and 23 days into online communications lies a tipping point where meeting offline becomes more disappointing than successful. Hyperpersonal communication may lead daters to create an idealised version of their communication partner in their mind, and differences between that ideal and reality may become greater the longer they communicate online without meeting. Hence, when daters find someone that they enjoy communicating with online, they should move to meeting offline within a reasonably short period of time.

9.8 Messaging in online dating

The platform that the dater uses can also impose self-presentation constraints. Traditional online dating profiles can be quite extensive in the amount of information revealed, whereas PNAs tend to have simple profile interfaces consisting of a photograph, basic information, and geographic location and lack the richness of dating profiles. But despite the disparities in profile type on different platforms, much of the self-presentation and impression formation occurs in private messaging on all platforms (Zytko et al., 2014). At this stage of contact, what daters feel is most lacking and would be highly desirable to have is feedback on the impressions they are creating. Men in particular tend to experiment with different types of message in their desire to get responses, and often develop anxiety and are hurt by the lack of response they receive (Zytko et al., 2014). This lack of response can be partly explained by differences in the quantity of messages received by women and men, where women (and also gay men) are often overwhelmed by a limited number of interesting, and a great number of generic messages that they must take time to sift through (Smith & Duggan, 2013). They may find it difficult to reply to all of the interesting messages, or these may get lost in the deluge. Providing information to both male and female daters about how people typically behave on dating sites may help men understand that it might not be their approach at fault, and may encourage women to take the initiative to seek out partners rather than wait to be approached. Many dating sites do not incorporate any aspects of community into their offering, meaning that most interaction happens in private messaging, and the formation of norms of behaviour in the site is difficult to manage. Masden and Edwards (2015) looked at the extensive online communities that have emerged on third party websites about specific dating sites. They found that these provided a great deal of information about the norms of the dating sites, brought many of the hidden aspects of dating behaviours (such as harassment) into the light and supported users in dealing with them, and helped users develop more successful strategies for their own dating experiences. People who had been successful in online dating would sometimes participate in these communities in order to help others achieve success, thus dating becomes a collaborative effort rather than a competitive one. We know that many people use dating sites not only to find partners, but also to be social and for entertainment (Clemens et al., 2015). The addition of a strong and vibrant online community within a dating site may increase participation by these and other users, and support users in overcoming some of the problems associated with dating sites.

9.9   Attraction, mate selection and dating experiences

At the heart of what daters are seeking is physical attractiveness and rapport with a partner (Couch & Liamputtong, 2008). The filtering process used to determine attraction and authenticity is important to daters and helps them feel more confident about the verity of a prospective partner. This process usually involves an escalation of contact from silent screening for physical attractiveness without contact, through browsing attractive profiles to seek disqualifying information such as smoking or drinking habits, to making contact with qualified daters through private messaging. If positive this leads to a transition to phone or email, meeting face-to-face for coffee or another low-investment meeting, and meeting for what is considered a romantic date (Couch & Liamputtong, 2008). Despite this time-consuming filtering system, where people spend substantially more time evaluating possible dates than actually meeting them, bad first dates are far more common than good ones, and people are not confident that they have formed accurate impressions of their communication partners (Frost et al., 2008; Zytko et al., 2014). This lack of confidence in being able to accurately perceive others' experiential characteristics, such as rapport, warmth, humour, or kindness on dating sites, and the feeling that others are not perceiving them as they wish to be because the dating sites limit their ability to convey complex experiential information about themselves, is one of the main frustrations for daters.

In fact most impressions formed online are found to be inaccurate when couples meet offline, often resulting in disappointment (Zytko et al., 2014). Dating sites enable partner searches by characteristics such as height, education, and marital status, which are not as important in actual relationship-building as experiential traits, while the important experiential traits are not searchable. Researchers in the field have proposed a number of solutions, including the addition of collaborative games or experiences that would allow interested matches to cooperate or interact in an environment or on a task. This would give potential couples a starting point for more realistic and experiential portrayals of themselves outside the static and inflexible realms of the dating profile and messaging system (Frost et al., 2008; Zytko et al., 2015). Gaming is already an increasingly popular way to meet romantic partners, though romance is not usually the primary purpose of the game (Hall, 2014). It is possible that a game-based dating site might be successful in overcoming the hurdles of impression formation on dating sites. Frost et al. (2008) tested the effect of virtual dates on online dating pairs, on likeability and intention to meet offline. The virtual date was a simple gallery environment with real-time messaging, where daters could access experiential information about the other person through more naturalistic and less controlled interaction. Daters who experienced the virtual date had greater liking of and intention to meet their partner than those who only read the dating profile of a potential partner. Kotlyar and Ariely (2013) also found that the use of avatars in communicating online could help improve relationship formation. Avatars that were in some way responsive and exuded social cues were perceived to be more positive than chat through text.

9.10   Matching and decision-making

A fundamental offering of online dating is the broad choice of options that are available. In many ways the marketing metaphor is an accurate one, and it is one that resonates with many daters (Heino et al., 2010). The design of dating sites encourages daters to employ shopping-like strategies towards dating, looking for the perfect product, rather than focussing on building quality relationships, and this has an effect on how daters perceive themselves and others and on how they behave. Often, desirable characteristics (or features) are listed as though from a catalogue of products, and the level of selectivity that daters employ may be calculated from their perception of their own level of desirability and the level of competition in the market (Heino et al., 2010). These strategies do not necessarily result in the best possible matches for daters. The filtering process encourages daters to search using individual or a selection of characteristics, rather than the common offline approach of filtering through a universal appraisal of a whole person. Instead of directing users to look at social interaction or rapport, dating sites focus on characteristics such as age and height and this leads daters to believe that these are important when they have little actual impact on the quality of a relationship (Fiore & Donath, 2005; Heino et al., 2010). Overall this may encourage daters to view themselves and others as commodities, which can be purchased, assessed, and discarded without consideration.

The matching systems and algorithms that dating sites use vary considerably, at least in how they are described by the sites. For example, Match.com appears to leave the searching for suitable matches up to the user, while a site like eHarmony.com has a very rigid system that requires a lot of information to be entered in a linear format before it matches based on personality. However, in recent years, Match and other sites like Plenty of Fish have also incorporated a lot of questions about personality into their profile creation process and use these as part of their matching process. Despite dating sites' apparent focus on personality for matching, research has found that actual similarity of personality between potential partners has little effect on attraction and the success of long-term relationships. However, perceived similarity does have some effect on attraction to strangers. In other words, we like people who we think have a similar personality to ourselves, but not necessarily people who actually do (Montoya et al., 2008). What dating sites actually use to determine possible matches is a combination of a dater's stated preferences, their actual behaviour on the site (which might not match their stated preferences as people often behave different to how they think they will), and *collaborative filtering*, where the preferences and behaviour of other similar users of the site are factored into account to predict an individual's preference (Bridle, 2014). The more information that a dater provides on their profile and the more they interact with the site, the more information the algorithm has available to identify possible matches.

The way in which you build your profile can also have an effect on how attractive you find the matches that the site presents you with. One finding from a study by Rao et al. (2009) found that users who entered their information while their personal profile picture was displayed beside the entry fields had increased self-awareness, and this had an effect on how they presented themselves. Additionally, those who had their profile picture displayed during this process also liked the recommendations that were returned to them more than those who didn't, despite the fact that deliberately poor results were returned. If a dating site doesn't offer this option, keeping a profile photograph visible during profile creation could lead to a more satisfying experience on the site. They also found that when poor results were presented to users, giving the option of providing immediate feedback on those results may lead to less frustration while a dating site algorithm is learning about a new user.

Having a vast number of options available at your fingertips can also result in bad decision-making processes. Wu and Chiou (2009) found that having more choice options available results in less selectivity by decision-makers. Selectivity is the allocation of more time to better options in the decision-making process. When the availability of too many options reduces selectivity, it results in daters being unable to screen out irrelevant or poor options. An excess of choice can also result in increased searching and cognitive overload for searchers, which in turn can result in mistakes being made in the search and poor choices being made.

9.11 Experiences of online dating

While daters generally report positive experiences of online dating, a sizeable number of women and a smaller number of men report harassment and uncomfortable experiences (Masden & Edwards, 2015; Smith & Duggan, 2013). Many users of PNAs like Grindr are almost entirely anonymous, and though most online daters present themselves positively but honestly (Ellison et al., 2006; Whitty, 2008), anonymity allows users to present their identity in any way they wish and to mask their intentions for using the sites. Gay male users of PNAs have experienced or witnessed discrimination of many kinds including against ethnicity, weight, health status, unattractiveness, and femininity (Van De Wiele & Tong, 2014). Gay male and bisexual users of PNAs have expressed concern over their privacy and personal safety while using applications like Grindr (Van De Wiele & Tong, 2014). Recent media coverage in a number of countries has highlighted cases where gay men have been targeted through dating sites and attacked by homophobes when they arrived to meet their date, so caution is certainly advised (for example, see Leogue, 2015). Dating sites frequently offer guidance on safety, with common tips such as not giving your home address and meeting in a public space in daytime for the first time. However, these tips tend to be located on information pages that people may not access, and it may be advisable to make them prominent within the messaging system or the sign-up process in order to increase safety awareness. This is perhaps particularly relevant for sites commonly used by the LGBT community who tend to be the target of these attacks, and in particular for sites like Grindr, on which users are more likely to arrange late-night, casual encounters.

9.12   How do relationships started online fare in comparison to offline ones?

Long-term outcomes for couples meeting online are generally positive. Hall (2014) found that couples who met through offline or through SNS and online dating had similar levels of marital satisfaction. Some study results on the success of long-term relationships are contradictory, with Cacioppo et al. (2013) finding that couples meeting online are likely to have slightly higher marital satisfaction, while Paul (2014) found that couples meeting online are less likely to marry and are a little more likely to break up than those who met offline. It should be noted that studies examining long-term outcomes of meeting online are evolving as quickly as the landscape of romantic relationships and results will possibly become clearer with time. Overall partnership rates have remained steady in the US over the years. However this is not the case for same-sex couples, whose partnership rates have risen significantly since the introduction of the Internet (Rosenfeld & Thomas, 2012). This may be due partly to an increasingly accepting society, but the effect of the Internet has undoubtedly had a seismic impact on LGBT relationships.

9.13 Conclusion

This chapter has addressed the rise of online dating and the impact that this has had on a wide variety of relationship-seekers. The Internet has fundamentally changed the fabric of the dating environment for many people. People choose to date online for many reasons, and while finding a relationship is a key motivator, it is not the only positive outcome of online dating. People also use it to find friends, network, and provide entertainment. The online dating environment changes the way in which we present ourselves and perceive others. It also has implications for how we choose potential mates and for the initial course of those relationships. While most daters have a positive experience dating online, there are also considerable frustrations encountered because of the limitations of the sites as they currently exist. Some recommendations for interactivity, community, and interface design improvements to dating sites have been suggested to overcome these frustrations and hurdles to meeting an ideal partner. This dating landscape is a rapidly evolving one and is not without drawbacks, but online dating ultimately enables millions of single people around the world to connect in ways that would have been impossible to conceive of two decades ago.

9.14 References

Aretz, W., Demuth, I., Schmidt, K., & Vierlein, J. (2010). Partner search in the digital age. Psychological characteristics of online-dating-service-users and its contribution to the explanation of different patterns of utilization. *Journal of Business and Media Psychology, 1,* 8–16.

Blackhart, G. C., Fitzpatrick, J., & Williamson, J. (2014). Dispositional factors predicting use of online dating sites and behaviors related to online dating. *Computers in Human Behavior, 33,* 113–118.

Bridle, J. (2014, February 9). The algorithm method: How internet dating became everyone's route to a perfect love match. *The Observer.* Retrieved April 9, 2015 from http://www.theguardian.com/observer

Cacioppo, J. T., Cacioppo, S., Gonzaga, G. C., Ogburn, E. L., & VanderWeele, T. J. (2013). Marital satisfaction and break-ups differ across on-line and off-line meeting venues. *Proceedings of the National Academy of Sciences, 110*(25), 10135–10140.

Clemens, C., Atkin, D., & Krishnan, A. (2015). The influence of biological and personality traits on gratifications obtained through online dating websites. *Computers in Human Behavior, 49,* 120–129.

Cloyd, L. (1977). Effect of acquaintanceship on accuracy of person perception. *Perceptual and Motor Skills, 44*(3), 819–826.

Couch, D., & Liamputtong, P. (2008). Online dating and mating: The use of the internet to meet sexual partners. *Qualitative Health Research, 18*(2), 268–279.

Downey, G., & Feldman, S. I. (1996). Implications of rejection sensitivity for intimate relationships. *Journal of Personality and Social Psychology, 70*(6), 1327.

Ellison, N. B., Hancock, J. T., & Toma, C. L. (2012). Profile as promise: A framework for conceptualizing veracity in online dating self-presentations. *New Media & Society, 14*(1), 45–62.

Ellison, N., Heino, R., & Gibbs, J. (2006). Managing impressions online: Self-presentation processes in the online dating environment. *Journal of Computer-Mediated Communication, 11*(2), 415–441.

Fiore, A. T., & Donath, J. S. (2005). Homophily in online dating: When do you like someone like yourself? *Computer-Human Interaction 2005*, 1371–1374.

Fiore, A. T., Taylor, L. S., Mendelsohn, G. A., & Hearst, M. A. (2008). Assessing attractiveness in online dating profiles. *Computer-Human Interaction 2008, 797*. New York, USA: ACM Press.

Fiore, A. T., Taylor, L. S., Zhong, X., Mendelsohn, G. A., & Cheshire, C. (2010). Who's right and who writes: People, profiles, contacts, and replies in online dating. *Proceedings of the 43rd Hawaii International Conference on System Sciences*, 1–10. CA, USA: IEEE Computer Society. doi:10.1109/HICSS.2010.444

Frost, J. H., Chance, Z., Norton, M. I., & Ariely, D. (2008). People are experience goods: Improving online dating with virtual dates. *Journal of Interactive Marketing, 22*(1), 51–61.

Fullwood, C. (2015). The role of personality in online self-presentation. In A. Attrill (Ed.), *Cyberpsychology*. Oxford: Oxford University Press.

Gibbs, J. L., Ellison, N. B., & Heino, R. D. (2006). Self-presentation in online personals: The role of anticipated future interaction, self-disclosure, and perceived success in Internet dating. *Communication Research, 33*(2), 1–26.

Goedel, W. C., & Duncan, D. T. (2015). Geosocial-networking app usage patterns of gay, bisexual, and other men who have sex with men: Survey among users of Grindr, a mobile dating app. *JMIR Public Health and Surveillance, 1*(1), e4.

Guadagno, R. E., Okdie, B. M., & Kruse, S. A. (2012). Dating deception: Gender, online dating, and exaggerated self-presentation. *Computers in Human Behavior, 28*(2), 642–647.

Hall, J. A. (2014). First comes social networking, then comes marriage? Characteristics of Americans married 2005–2012 who met through social networking sites. *Cyberpsychology, Behavior, and Social Networking, 17*(5), 322–326.

Heino, R. D., Ellison, N. B., & Gibbs, J. L. (2010). Relationshopping: Investigating the market metaphor in online dating. *Journal of Social and Personal Relationships, 27*(4), 427–447.

Henry-Waring, M., & Barraket, J. (2008). Dating and intimacy in the 21st century: The use of online dating sites in Australia. *International Journal of Emerging Technologies and Society, 6*(1), 14–33.

Higgins, E. T. (1987). Self-discrepancy: A theory relating self and affect. *Psychological Review, 94*(3), 319.

Hitsch, G. J., Hortaçsu, A., & Ariely, D. (2010). What makes you click? Mate preferences in online dating. *Quantitative Marketing and Economics, 8*(4), 393–427.

Hogan, B., Li, N., & Dutton, W. H. (2011). A global shift in the social relationships of networked individuals: Meeting and dating online comes of age. Oxford Internet Institute, University of Oxford. Retrieved from http://www.oii.ox.ac.uk/research/projects/?id=47

Kotlyar, I., & Ariely, D. (2013). The effect of nonverbal cues on relationship formation. *Computers in Human Behavior, 29*(3), 544–551.

Lee, A. Y., & Bruckman, A. S. (2007, November). Judging you by the company you keep: Dating on social networking sites. In *Proceedings of the 2007 International ACM Conference on Supporting Group Work* (pp. 371–378). Florida, USA: ACM.

Lo, S. K., Hsieh, A. Y., & Chiu, Y. P. (2013). Contradictory deceptive behavior in online dating. *Computers in Human Behavior, 29*(4), 1755–1762.

Leogue, J. (2015, February 18). Warning over homophobic 'catfish' attacks. *Irish Examiner*. Retrieved from http://www.irishexaminer.com

Mascaro, C. M., Magee, R. M., & Goggins, S. P. (2012, February). Not just a wink and smile: an analysis of user-defined success in online dating. In *Proceedings of the 2012 iConference* (pp. 200–206). New York, USA: ACM.

Masden, C., & Edwards, W. K. (2015, April). Understanding the role of community in online dating. In *Proceedings of the 33rd annual ACM conference on human factors in computing systems* (pp. 535–544). New York, USA: ACM.

Montoya, R. M., Horton, R. S., & Kirchner, J. (2008). Is actual similarity necessary for attraction? A meta-analysis of actual and perceived similarity. *Journal of Social and Personal Relationships, 25*(6), 889–922.

Morgan, E. M., Richards, T. C., & VanNess, E. M. (2010). Comparing narratives of personal and preferred partner characteristics in online dating advertisements. *Computers in Human Behavior, 26*(5), 883–888.

Norton, M., Frost, J., & Ariely, D. (2007). Less is more: The lure of ambiguity, or why familiarity breeds contempt. *Journal of Personality and Social Psychology, 92*(1), 97–105.

Paul, A. (2014). Is online better than offline for meeting partners? Depends: Are you looking to marry or to date? *Cyberpsychology, Behavior, and Social Networking, 17*(10), 664–667.

Ramirez, A., Fleuriet, C., & Cole, M. (2015). When online dating partners meet offline: The effect of modality switching on relational communication between online daters. *Journal of Computer-Mediated Communication, 20*(1), 99–114.

Rao, S., Hurlbutt, T., Nass, C., & JanakiRam, N. (2009, April). My dating site thinks I'm a loser: Effects of personal photos and presentation intervals on perceptions of recommender systems. In *Proceedings of the SIGCHI Conference on Human Factors in Computing Systems* (pp. 221–224). New York, USA: ACM.

Rosenfeld, M. J. (2010). Meeting online: The rise of the internet as a social intermediary. *Paper presented at the annual meeting of the American Sociological Association annual meeting*. Retrieved from http://www.allacademic.com/meta/p409508_index.html

Rosenfeld, M. J., & Thomas, R. J. (2012). Searching for a mate the rise of the Internet as a social intermediary. *American Sociological Review, 77*(4), 523–547.

Smith, A., & Duggan, M. (2013). Online dating & relationships. *PEW Internet & American Life Project*. Washington, D.C., US. Retrieved November 10, 2013 from http://www.pewinternet.org/Reports/2013/Online-Dating.aspx

Suler, J. (2004). The online disinhibition effect. *CyberPsychology & Behavior, 7*(3), 321–326.

Valkenburg, P. M., & Peter, J. (2007). Who visits online dating sites? Exploring some characteristics of online daters. *CyberPsychology & Behavior, 10*(6), 849–852.

Van De Wiele, C., & Tong, S. T. (2014, September). Breaking boundaries: the uses & gratifications of Grindr. In *Proceedings of the 2014 ACM international joint conference on pervasive and ubiquitous computing* (pp. 619–630). New York, USA: ACM.

Walther, J. B. (1996). Computer-mediated communication impersonal, interpersonal, and hyperpersonal interaction. *Communication Research, 23*(1), 3–43.

Whitty, M. T. (2008). Revealing the 'real' me, searching for the 'actual' you: Presentations of self on an internet dating site. *Computers in Human Behavior, 24*(4), 1707–1723.

Wu, P. L., & Chiou, W. B. (2009). More options lead to more searching and worse choices in finding partners for romantic relationships online: An experimental study. *CyberPsychology & Behavior, 12*(3), 315–318.

Zytko, D., Freeman, G., Grandhi, S. A., Herring, S. C., & Jones, Q. G. (2015, February). Enhancing evaluation of potential dates online through paired collaborative activities. In *Proceedings of the 18th ACM conference on computer supported cooperative work & social computing* (pp. 1849–1859). New York, USA: ACM.

Zytko, D., Grandhi, S. A., & Jones, Q. (2014, November). Impression management struggles in online dating. In *Proceedings of the 18th international conference on supporting group work* (pp. 53–62). New York, USA: ACM.

10

Online Consumer Behaviour

Nicola Derrer-Rendall and Alison Attrill

10.1 Introduction

In westernised cultures, there is an advancing shift from offline to online activities for many routine behaviours, including shopping and banking (Statista, 2015). Additionally, it was reported in 2014 that 28.3% of all UK fashion purchases and 28.8% of general goods purchases were made online. With online retail equalling big money, it is no wonder that interest is mounting in the area of online consumer behaviour. Increases in online consumerism may in part be due to the availability of online shopping via smartphones and tablets around the clock, anytime, anywhere. Online commerce and shopping behaviour are continuously evolving, to the point where some even blame the Internet for the demise of traditional high street shopping here in the UK. With this shift comes a need for retailers to consider the psychological aspects of online retail in the development of websites, along with the attraction and retention of online shoppers in order to provide a consistently positive consumer experience. Drawing on theories and research from cyberpsychology, and consumer, social, and cognitive psychology, this chapter will highlight some of the key variations between consumer experiences in online and offline shopping, and will consider ways in which retailers are attempting to blur physical and virtual shopping experiences. We will also identify some of the key issues regarding satisfaction, trust, and loyalty in consumers' online shopping behaviours, and the impact of Electronic Word of Mouth and consumer complaint procedures. Some of the positives and negatives of online consumer experiences will also be touched upon, such as shopping compulsions and the opportunity of inclusion and convenience provided by online shopping. A good place for us to begin this discussion is to consider how and why online retail might be gaining preference over offline shopping.

10.2   Online versus offline shopping

A common perception is that shopping online is generally quicker, easier, and more convenient than traditional offline shopping. This is possibly a somewhat idealistic viewpoint with no supporting consensus within the academic literature. Whilst some consumers may prefer or feel more comfortable going into a physical store in a more traditional offline shopping space, others may prefer to shop online. There are many characteristics about both the shopping experience and consumers that might enlighten us as to why people might choose one over the other. Imagine for instance that you need a new bed. Where do you begin your investigations as to which bed to buy? Do you go to your local retail centre and visit only available bed shops, or do you travel wider afield to compare the same make of bed across a number of physical stores? Alternatively, you might already have looked online to narrow down the choices of the bed and mattress you would like prior to visiting offline retailers in search of the best deal. This choice might have been influenced by online reviews and consumer reports. Conversely, you might have looked at beds in offline shops and then sought out the best deal online. The point is that there are numerous ways in which our online and offline shopping behaviours combine or remain entirely separate from one another. For retailers, a minefield of individual characteristics, preferences, perceptions, and interpretations ensues that needs to be translated into financial profit from both online and offline consumer experiences. When taking all of these factors into consideration, it becomes clear why some large retail chains have gone into liquidation in recent times. Offline retailers need physical spaces that cost a lot of money to display and store items. They also require more staff than online retailers. Of course, online retailers need to factor distribution and delivery into account, but their low storage and website hosting costs allow a high volume turnover for small unit profit. Many retailers are now offering combinations of online and offline trading, but traditional high street retailers struggle to compete with companies that operate solely online, especially if they are working on a high unit profit with low volume turnover. One way in which they might translate online footfall into profit is to understand individual differences in the ways in which people shop online and offline.

10.3   Individual differences

There are some well-known individual factors that influence shopping preferences and choices. For example, Passyn et al. (2011) identified significant differences between male and female opinions about the benefits and problems when purchasing online compared to offline shopping. Women identified an inability to examine the item, the experience being "boring," and the cost of postage as significant issues that were not highlighted by males as concerning. The authors suggest that these negative opinions for females are associated with a lack of shopping experience where the purchase is made online. In contrast, male consumers' negative opinions relate to the actual transactional element of the process. They perceive the completion of the purchase rather than the shopping experience to be negative. Both sexes, however, identified the flexibility to be able to shop anywhere at any time online as a positive. Passyn et al. concluded from this that *"women shop to shop while men shop to buy"* (p. 104). It would thus appear that the shopping experience resembles a social activity for women, whereas males see shopping as a functional process to achieve a specific outcome.

Further individual differences were highlighted in Lieber and Syverton's (2012) observation of variations in age, education, household income, and ethnicity regarding general Internet use. Older people (60+ years), those with lower level educational qualifications, and those from lower income households used the Internet less frequently. These findings appear to be mirrored in some patterns of online purchasing behaviours. According to the Office for National Statistics (ONS) (ons.gov.uk, 2014) 74% of all adults in the UK bought goods or services online in 2014, an increase from 53% in 2008. This increase is reflected across the lifespan. For instance, 90% of 25- to 34-year-olds reported making an online purchase in 2014 and consumers aged 65 and over demonstrated an increase from 16% in 2008 to 40% in 2014. This development implies that although the 65+ population are less likely to use the Internet than younger age groups, older users are possibly becoming more *au fait* with the online world and therefore more willing to engage in online shopping. As we know, however, from Chapter 4, different age groups use the Internet, with different aims and goals. Thus, an 18-year-old might go online to purchase or research different items from those desired by a 50-, 60- or 70-year-old. The implication for online retailers with this increase in older consumers shopping online does suggest a need for retailers to consider their target audiences. For instance, online store layouts might be created with specific target audiences in mind. If aimed at older consumers, different features might be required from a store aimed at under 30-year-olds. These features could range from the font, font size, and colour used to the way in which hyperlinks are embedded, and how payments are taken. But what is it that people buy online? It would appear that clothing was the most popular of online purchases in 2014, with 49% of all adults making an online clothing purchase. More than half of all females (52%) bought clothing online in 2014 compared to only 46% of males. This is contrary to our earlier notion that women enjoy the actual physical shopping experience (Passyn et al., 2011). Unfortunately, the offline purchasing habits from the same participants are not reported. It would be interesting to ascertain whether these online shopping statistics replace offline shopping habits, or are in addition to offline shopping. There are of course many advantages to be had for women shopping for clothing online, most notably trying on the clothes in the comfort of their own homes. The question that would arise from this is whether this positive sufficiently overrides the possible negatives of waiting for delivery and maybe returning items. In order to try to explore this further, it is useful to consider some of the work relating to the shopping experience itself.

10.4   The shopping experience

One of the fundamental factors that might play a role in influencing a choice between online and offline retail is one's previous shopping experience(s) for desired or similar items. Let's return to the example of buying a bed. In addition to the factors previously outlined, other factors that will likely influence your choice of online or offline purchase include where you may have previously purchased a bed, the nature of the sales staff (were they friendly, knowledgeable, and helpful?), and the range of beds to choose from. If you are purchasing the bed online, the considerations will likely be different. One of the over-arching factors in this decision might be whether you can get a real feel for how the item actually looks and feels based on your offline or previous experiences. One of the reasons for this could be humans' desire to choose the familiar. We know from social psychological principles that humans like items, objects, and people who are familiar (e.g., Zajonc's (1968) mere exposure effect). This could also apply to the actual mode (online/offline) of shopping. It may be the case that shoppers opt for their most used mode of retail because it is familiar. This familiarity can be created far more easily offline than online with retailers, for example, creating a shopping experience that utilises shoppers' senses of smells and sound that are not necessarily achievable online.

The offline shopping experience can thus have positive effects on consumers' buying behaviours, from the music, lighting, and scent of the retail environment (Grewal et al., 2014) to being able to sample merchandise. Whether trying on clothes, sampling food or just checking on the functionality and usability of an item of interest (Spence et al., 2014), the recreation of sensual experiences online is more difficult. It is this that might provide one of the biggest differences between online and offline shopping experiences. When shopping online the senses may be less engaged, making it less likely that shoppers will be seduced into purchases, as may happen in the offline shop. That said, maybe there are other factors that can more easily be replicated or created online. Findings from Puccinelli et al. (2009) suggest, for instance, that a positive in-store purchasing experience can cement the future relationship the consumer has with the retailer in both offline *and* online purchases. Visual Merchandising Design (VMD), according to Law et al. (2012), involves both the interior and exterior of the retail store, and within these two environments there are important features, such as the combinations of colours utilised, mannequin and props selected, as well as fixtures and fittings, which when combined effectively enhance the shopping experience for the consumer (Babin et al., 2003), encouraging them to stay longer and spend more (Donovan et al., 1994). These features can, however, also evoke the opposite impact. When a consumer is exposed to overly loud music, very bright lighting and inaccessible shelving, for instance, a negative emotional reaction might occur. This then leads to less time spent in store and no purchase being made (Donovan & Rossiter, 1982). At this point, the consumer might turn to the Internet to make the purchase from the comfort of their own home. It is difficult for the non-physical e-store to wholly replicate either the positive or negative offline experience of shopping, but there are some aspects of online shopping which remain important, such as visual layout.

Research identifying that online VMD can have a positive impact on consumers' satisfaction and intentions to buy (Park et al., 2005; Wu et al., 2008) has led to the exploration of further important features in online retail. One such feature is that of the impulse purchase. Consider your own shopping behaviour: when are you most likely to make an impulse purchase? If doing so offline, does the act of picking up the item and physically handing over cash provide a positive or negative impact on your mood when making the impulse buy? Thanks to the mass media we are now all familiar with the term shopaholic, but what does this mean in relation to online versus offline shopping? Offline, handing over cash that a consumer knows they need to pay bills could curb shopping behaviour. Online, no physical transaction is required. People can run up debt without ever facing a person on the other side of a check-out and thus possibly avoiding any notions of guilt or psychological stress. Impulse purchases are associated with a rush of adrenaline – almost like an addict receiving a drug or alcohol, it is a rush of positive emotions and feelings that may at some point subside into guilt, concern, or worry about having spent money needed elsewhere. At what point would these feelings kick in online? It might be the case that the positive emotions experienced when purchasing an item subside by the time the item arrives. Alternatively, that rush might not happen until the items arrive, and as no physical money changes hands, the consumer might not experience the negative emotions of impulse buying. Understanding the interplay of the psychological factors involved in both compulsive and impulsive shopping could therefore be extremely useful to online retailers for the ways in which they conduct business online.

10.5   Risk and familiarity

One way in which a rush might be achieved when shopping online is through risk. Choi and Lee (2003) reported that consumers perceive a high level of risk with some types of online shopping, such as for clothing. The consumer is unable to see the true colour, feel the material, or try the clothing on. Therefore, the risk involved in is far greater than being in store. That said, retailers are working very hard to make the online shopping experience as close to the offline experience as possible. Online clothes stores often use real life models in videos to demonstrate clothing along with zoom-able pictures from various angles. Indeed, such an approach to advertising clothes online has been found to have a positive impact on consumers' emotional reactions to products and their intention to purchase these online (e.g., Kim et al., 2009). This ability to see the product on a model (rather than as a flat item on a plain background) enables the consumer to make what they perceive to be a more effective judgement of the suitability of the item for their purpose. It also removes some of the perceived risk associated with online purchases. These applications extend beyond the buying of clothes online. When looking for furniture online, some stores now provide apps for mobile phones and tablets which can be used to superimpose a piece of furniture into a picture of one's current room. The same sort of app can be used for testing out paint colours and even planning whole rooms such as kitchens and bathrooms. The bottom line is that consumers increasingly rely on digital tools when shopping online. These tools are often born out of some consideration of the psychological features associated with selling items online, such as the aforementioned notion of familiarity: by placing a considered item into a live picture of one's own home, for instance, it can take on a notion of familiarity. But what happens when these factors feed into a continued use of online retail, above and beyond what one can actually afford?

10.6 Online consumer behaviours

Thus far we have considered online retail from a retailer perspective. The ease with which people can browse and shop online anywhere at any time via a multitude of gadgetry provides potential for over-spending and the development of compulsive shopping behaviours. Greenfield (1999) identifies several factors which contribute to the addictive nature of the Internet, including access and availability, disinhibition, reward and reinforcement, and boundaries.

10.6.1 Access/availability

Many people now have smartphones and tablets with them round the clock, even when they are sleeping. They are no longer constrained by retail opening hours and can shop online 24 hours a day, seven days a week, and 365 days a year, providing the opportunity for unlimited shopping.

10.6.2 Disinhibition

This notion is similar to the online disinhibition effect outlined by Suler (2004). Online, shoppers may experience a perceived anonymity that enables them to be less inhibited, or less likely to shop within the restrictions experienced in offline shopping. It is, however, questionable whether this perceived anonymity is really that different to offline shopping nowadays. The magnitude of retail parks could make offline shopping equally anonymous. Disinhibition might, however, play a larger role online because people feel freer and less confined by moral judgements than when shopping offline. If one wishes, for instance, to purchase items such as sex-toys, which could evoke embarrassment in an offline store, the perceived disinhibition and anonymity could be an online shopper's advantage. This idea is linked to Greenfield's (1999) notion of boundaries.

10.6.3 Boundaries

Unlike the physical environment, one could argue that both situational and dispositional boundaries are reduced, if not eradicated online.

The online shopping experience is endless and therefore enables the shopper to become consumed by browsing from one store to another, possibly without exhausting all available options on offer. Additionally, they might go online with a sense of what they are looking for, but be led more and more astray in what they actually end up purchasing by these endless options. It might thus be the case that whilst offline stores utilise tools and tricks to lure people into parting with their money by speaking to more of the human senses, online it is the variety and endless shopping experience that entices people to spend money.

10.6.4   Reward/reinforcement

Social psychology informs us that rewards reinforce behaviour (Skinner, 1938). If online shopping reinforces the pleasure gained from finding a desired item or a good deal, continued shopping might be sought to replicate that positive feeling. As with most compulsive behaviours, however, the reward may become less and less stimulating over time. Extensive time may then be spent seeking further rewards and reinforcement. These could become unpredictable and intermittent, encouraging some consumers to spend extensive amounts of time searching and browsing. Thus, the negative impact of continued online shopping may not necessarily be one of monetary value, but could also be linked to the time spent on online shopping. Browsing during work hours may, for example, lead to disciplinary action, but does this type of behaviour always morph into an online shopping addiction?

10.7   Compulsive online shopping

Given that there is no blood-brain barrier transfer of a substance, which is characteristic of addiction, many authors prefer to use the word compulsion for non-substance-related behaviours. The term Online Shopping Addiction is a relatively new concept (see Rose & Dhandayudham, 2014). Older research does, however, highlight self-regulation as a potential attribute which is lacking in compulsive online shoppers (La Rose & Eastin, 2002). Retailers have several components to their websites which should encourage self-regulation, such as the shopping basket showing the consumer exactly what they have bought and the total running cost, prior to proceeding to check-out and purchase. The previous purchase history facility should remind the consumer what they have previously purchased, and when and how much they have spent, and encourage them to self-regulate their spending behaviour. Both of these aspects of the shopping experience should encourage self-regulation, but in a compulsive consumer, these have limited impact on decision-making, especially if all of their spending does not occur on a single website. Indeed, it might be argued that retailers themselves could be actively destroying a consumer's self-regulation through their strategically placed adverts and time-limited special offers.

More recently, Rose and Dhandayudham (2014) identified several potential predictors of compulsive shopping behaviour, suggesting that some factors are very much embedded in the general addiction characteristics of individuals, such as low self-esteem, negative emotional state, and low self-regulation. Other predictors are not as generalisable to other addictions, such as gender (females are more likely to experience addictive buying behaviour), enjoyment, and the very specific online factor of social anonymity (the ability to keep completely private one's buying behaviour). One might question why online retailers would be invested in stopping people parting with their money. It would seem that there is more of a focus on advertising with pop-ups, flash sales, and discount vouchers for regular custom, all serving to fuel compulsive shopping further through the enjoyment experienced from the perceived bargains and sale prices encountered. Unlike the makers and distributors of alcohol and cigarettes, the online retail community are at present unaccountable for the negative consequences of continued spending online.

As an aside but related comment to this section on compulsive shopping, the reader might also consider whether we should perceive online betting and gambling facilities, donation websites, and the apparent ease with which consumers can attain credit online as negatively capitalising on the psychology of online retail. These include online bingo websites, sexual webcam websites, and other sites that operate solely with the intention of making a profit. It is worth considering whether these companies would be as sustainable offline as online, or whether it is the simple availability of the online sphere that cultivates the development of these behaviours. The same psychological principles of advertising and marketing might apply to attracting consumers to these websites, but there could be a far greater component of perceived risk, disinhibition, and anonymity involved in these rather than online retail websites. Should the reader wish to explore the currently available literature in this area further, Professor Azy Barak's website (http://construct.haifa.ac.il/~azy/refindx.htm) offers an up-to-date, extensive publications list on most aspects of online behaviour.

10.8   Consumer satisfaction and trust

One risk identified in online purchases is that of the importance of being able to return unwanted or faulty items. A stress-free returns process could increase trust in and satisfaction with both retailer and products. The Expectation-Confirmation Theory (Oliver, 1977, 1980) conceptualises the processes by which a consumer matches their expectations of a product or service to their post-service/purchase experience. This then develops the level of satisfaction the consumer has with their experience, the product, or the service received. The theory consists of four constructs: expectations (what consumers expect the product, service, or experience to be like), perceived performance (consumers' perceptions of the actual product, service, or experience – directly affected by original expectations), and disconfirmation of beliefs (consumers' evaluation of the product, service, or experience). When the product, service, or experience meets or exceeds the consumers' original expectation, a *positive* disconfirmation of beliefs is produced. If the evaluation falls short of original expectations, a *negative* disconfirmation of beliefs ensues. The final construct is satisfaction (the level of satisfaction and contentment consumers feel with the product, service, or experience). Consumers' satisfaction is influenced by their expectations (indirectly), perceived performance (indirectly), and the disconfirmation of beliefs (directly) held about the product, service, or experience. Imagine for example that you have booked a weekend away to Paris with a trusted travel agency that you have used several times before, and that you are going to stay in a hotel that has excellent reviews on several review websites. Your expectation would be of an idyllic, clean, and well-proportioned Parisian hotel that is well positioned to access all attractions. The actual experience you have of the holiday (perceived performance) may be that you enjoyed Paris, but your flight was delayed and you arrived late. The hotel was having building work completed, making it noisy, with reduced services and available facilities. Whilst the room and the food were good, they fell short of the rave reviews. Your expectations of the services and the experience did not accurately match your experience (negative disconfirmation of beliefs) and therefore you did not feel completely satisfied with the service and experience (satisfaction). The four factors thus combine to influence your likelihood of booking with the travel agent in the future, or recommending this holiday to others. The satisfaction that consumers have with products, service, or experience will thus impact on future loyalties to a retailer.

Given the range and number of retailers and websites that offer similar products to one another, retailers need to provide a satisfactory service if they wish to attract repeat business (see e.g. Bhattacherjee, 2001). The more satisfied the consumer is with the product, service, or experience, the greater the likelihood that they will re-visit/re-purchase from the retailer in future. In practical terms, the retailer may therefore have a single opportunity to create a loyal customer relationship through their service and customer experience during an initial purchase or visit to their online store. Returning to Social Psychology and considering cognitive psychological features, the primacy effect (the first impression we have) is known to play a significant role in our impression-formation and beliefs, whilst the recency effect provides the lasting impression we hold from a recent experience (Ebbinghaus, 1913). By making a good first impression the retailer can promote a positive impression and create in the consumer a positive attitude towards future shopping experiences with that store. We can delve further into social cognitive psychological aspects of association biases to suggest the emergence of a halo effect from positive online shopping experiences. That is, a cognitive bias that emerges when one positive experience is projected onto viewing all other aspects of (in this case) the retailer as equally positive (Thorndike, 1920). Unfortunately, these are not factors that online retailers alone capitalise on when thinking of website design and their services offered. Criminals are equally able to use the psychology of online retail in order to sell fake or stolen goods, or in some cases simply to defraud individuals without providing purchased items at all. Trust therefore becomes an essential component for reliable retailers in creating positive retailer-consumer relationships, and may be built up over repeated positive and satisfactory shopping experiences with an online retailer. In a study within the travel industry by Shankar et al. (2003), for instance, consumers' satisfaction was shown to be similar online to offline, but with loyalty to the company being higher when reservations were made online rather than offline. The authors suggest that this counter-intuitive finding could be due to the fact that the bookings made for hotels could be bookmarked easily online and therefore accessed far more easily to rebook in future.

McKnight et al. (2002) presented the Web Trust Model, which identifies the specific components of interactions that develop trust between retailers and consumers. The consumer's perception of the online retailer's attributes (trusting beliefs) lead to the consumer developing trusting intentions, which influence their trust-related behaviours. The trusting beliefs develop from satisfying experiences with the retailer. These encourage the consumer to believe that the product or service will be appropriate, which then leads to intentional and actual purchasing behaviour. Understanding the psychological components of building consumers' trust beliefs is vital for retailers, from designing websites to gaining site traffic and retaining customers. A retailer might, for example, garner trust by offering a first-time shoppers discount. If delivery of goods and item satisfaction subsequently ensue, the retailer will have created a good primacy and recency effect for the consumer along with an excellent online shopping experience. From the very outset then, successful retailers need to pay attention to psychological factors such as trust. If consumer trust is not built up, a retailer's reputation could be irrevocably damaged, not just for the single unhappy customer, but for the wider shopping society. That said, customer satisfaction is not based entirely on trust. Trust comes in many situational shapes and forms, including trust in the brand (company), technology, broadband providers, and financial institutions to keep our details safe. Dispositional features of trust are personality traits inherent to individuals and somewhat more difficult to generalise to either online or offline shopping behaviours. These are often intertwined with privacy concerns (e.g., will a company really keep personal details confidential or will they sell them to third party providers?). Understanding the role(s) of trust and privacy concerns in online retail is far more complicated than can be explored within the confines of this chapter. The reader is, however, directed to some work by Joinson and colleagues (e.g., Joinson et al., 2010), which provides a good starting point for exploring these factors further.

10.9   Consumer complaints

When purchases don't go to plan or products do not meet expectations, consumers' experience of the way in which the retailer deals with their issues and complaints can impact on the loyalty the consumer has to them in the future. If you have recently been unhappy with an online purchase, what did you do? Did you email the retailer or provide/post a negative review? Retailers are now faced with public naming and shaming, from stars awarded for feedback on retail sites to sites created merely for the purpose of writing reviews. Understanding when consumers will immediately provide negative feedback and how to deal with complaints to avoid such negativity could be crucial to the sustainability of an online shop. But is this really any different to offline sales and marketing? Consumer Complaining Behaviour (CCB) is experienced and dealt with quite differently in an online environment compared to face-to face (FtF) in-store interactions. Lee and Cude

(2012) highlight that online complaints enable consumers to express their dissatisfaction with a purchase or experience in a cost-effective way, allowing them to comment at a time convenient to them at little financial cost, without having to travel or deal with FtF confrontation. In their study of 480 undergraduate students, they highlight two main findings from the choices made about how and where to make a complaint. If a consumer made an online purchase, they were more likely to complain online. There was also evidence that the greater the dissatisfaction the more likely a consumer was to complain online. Perceived anonymity might also lead to a higher level of complaining online than offline. Many of us may feel that a complaint online is somehow easier to make rather than having to speak to someone over the telephone or FtF. Even though a retailer will know the personal details of the buyer, the interaction remains somewhat impersonal online and a buyer is not faced with facial cues or expressions, or the body language of the store attendant. Online, the customer services individual becomes a faceless person at the end of a digital connection, or a faceless voice at the end of a phone line. This is in line with Suler's (2004) aforementioned notion of online disinhibition, a phenomenon characterised by six factors, of which two seem pertinent when reviewing CCB. Online we may feel more able to voice our opinions in a controlled and timed manner, rather than having to respond to an individual's FtF and bodily reactions under conditions of relative anonymity (dissociative anonymity). We may also feel that we are more equal to the recipient of our complaint in terms of status and authority when sharing opinions online (minimisation of status and authority). These are important factors that help identify why consumers feel more comfortable sharing their dissatisfaction online rather than offline. This could, however, be a double-edged sword: given the wide-reaching realms of the Internet, a retailer could think that the loss of a single customer is irrelevant to their sustainability. Alternatively, they might be invested in avoiding public complaints at all costs by offering interactive services to reduce the likelihood of complaints. Ward and Ostram (2006) highlight that a shift in CCB to a very public online arena means that the times of being dissatisfied with the consumer experience or a product, and sharing this opinion with close friends and family, have made way for a far more public display of disappointment or dissatisfaction. As explained by Tripp and Grégoire (2011), online consumers often complain because they have experienced what is called double deviation. That is, they have first had a negative consumer experience and after attempting to rectify this situation, have experienced a poor or inadequate resolution.

The consumer perceives themselves to have been "let down" more than once by the same company. The double deviation in the eyes of the consumer is usually very serious and the consequences are therefore something that they feel unable to ignore (e.g., injury or financial loss). They experience a sense of real betrayal by the organisation, rather than just feeling unhappy or perturbed by a minor incident. The consumer may subsequently feel vindicated in sharing their negative thoughts online, regardless of the ensuing negative effect on an organisation's reputation. It thus appears essential that the retailer handles complaints effectively and efficiently, with excellent customer service, to avoid this double deviation potential and negative feeedback. Of course, if an online retailer has fewer staff than offline retailers, finding the manpower to deal with initial complaints may not always be easy. Many online companies therefore now offer cyberbots to deal with initial customer enquiries of any sort. These are the "people" who almost instantly pop up in a new window to ask if they can assist when visiting given websites. It would seem that companies feel that providing an almost FtF interaction online (even if the cyberbot often resembles a cartoon character) reinstates some of the situational and dispositional cues inherent in offline sales interactions. Much scepticism does, however, exist nowadays as to the authenticity of such cyberbots, and their influence on shoppers' buying decisions and site feedback.

We have painted a somewhat negative picture of the online consumer complaints culture. Consumers' sharing of their opinions is not always negative or detrimental to a retailer. There are ample examples of how people share positive experiences and reviews online. Nonetheless, it is human nature to focus on the negative, with people often having a tendency to retrieve negative over positive information, even when there are positive examples on offer. There are a number of memory-related features, such as retrieval-induced forgetting, directed forgetting, and active forgetting, that demonstrate how and why people are more prone to retrieve some types of information over others (see Anderson & Bjork, 1994; Anderson et al., 1994; MacLeod, 1975). The acceptance of negative over positive reviews and feedback could thus be individually determined by memory habits, but might also stem from cue and/ or association biases (e.g., McGeoch, 1936, 1942). Let's consider buying a car. If you had a certain type of car for many years, your memory of that car might be positive. However, if a relationship failed and over time many of the arguments that led to the break up took place in the car, this might create negative associations with that car. By association, all other cars made by the same company might now take on negative connotations. When browsing for a new car online, not only do your negative memories of your old car override any positive reviews posted, but they also taint views of any associated cars. Retrieval-induced forgetting (RIF) works, for example, when repeatedly retrieving one feature or memory leads to the suppression of associated memories. Constant retrieval of negative car memories thus leads to the active suppression of related positive memories. You would need to work consciously to push down those memories in favour of the positive ones. One way that this repeated retrieval may work online is via the *word of mouth phenomenon*. Online decision-making behaviour in relation to online retail has been linked with the impact of electronic Word of Mouth (eWoM). This can take the form of consumers actively searching for reviews and feedback on independent review websites, or from information provided on the chosen retailer's website. The impact that eWoM has on the consumer varies from research study to research study. Opposing the RIF notion above, some findings suggest that eWoM is most influential in situations where consumers are buying something of which they have limited knowledge, or when buying a new product on the market (Cheunga & Thadanib, 2012). Given the potential consequences of negative eWoM, the change in consumers' power and the impact of eWoM cannot be underestimated by retailers.

10.10   Inclusivity

Let us end this chapter with a positive take on online consumerism. There are many individuals for whom online shopping provides the only option for making choices and decisions about what they buy and where to buy it, such as people with physical or learning disabilities, or mental health/psychological issues (depression, agoraphobia, claustrophobia). There is a multitude of people who, for a whole array of reasons, are physically unable to shop offline. Being able to shop online could provide them with a sense of independence otherwise not present in their lives. Consumers with physical disabilities have highlighted interesting explanations for their use of online retailers in a study conducted by Childers et al. (2009), including that of convenience. Online shopping can eliminate the necessity to navigate around shopping centres and malls for consumers with physical disabilities. The use of online retailers by those dealing with mental health issues has been reported as being a safe, efficient, more confident, and comfortable experience (Caplan, 2003). The consumer can avoid psychosocially distressing situations by avoiding unplanned and uncontrolled social interactions.

This avoidance of anxiety-provoking situations is not always effective in the long term, however the ability to shop online without all of the fears and anxieties associated with social interactions and uncontrolled environments can enable those suffering with mental health issues to have more control and make reasoned choices in a safe and controlled environment.

10.11 Concluding comments

Throughout this chapter various features of online retail have been considered. We have briefly touched upon some of the social and cognitive (memory) aspects of consumers' online shopping experiences, whilst exploring individual and situational factors in more detail. The satisfaction and trust that consumers experience when shopping online has been shown to be influential on their intentions to re-purchase in the future, to complain, and to review and share online their positive or negative opinions with other potential consumers. It would appear that retailers should not be complacent about the potential power and influence that consumers have in the online world; nor should the online consumer be naïve about the intentions of retailers' use of their psychological understandings to reel buyers in to continue spending. Maybe, rather than considering online and offline retail experiences as distinct from one another in the future, companies might work towards a hybrid customer interaction that focuses on the overall consumer experience. A prime example of this hybrid interaction is a consumer looking at products in the physical environment whilst simultaneously using their smart phone to search for the best price or reviews of the product online.

10.12 References

Anderson, M. C., & Bjork, R. A. (1994). Mechanisms of inhibition in long-term memory: A new taxonomy. In D. Dagenbach & T. Carr (Eds.), *Inhibitory processes in attention, memory, and language* (pp. 265–325). San Diego, CA: Academic Press.

Anderson, M. C., Bjork, R. A., & Bjork, E. L. (1994). Remembering can cause forgetting: Retrieval dynamics in long-term memory. *Journal of Experimental Psychology: Learning, Memory & Cognition, 20*, 1063–1087.

Babin, B., Hardesty, D., & Suter, T. (2003). Color and shopping intentions: The intervening effect of price fairness and perceived effect. *Journal of Business Research, 56*(7), 541–551.

Bhattacherjee, A. (2001). An empirical analysis of the antecedents of electronic commerce service continuance. *Decision Support Systems, 32*(2), 201–214.

Caplan, S. E. (2003). Preference for online social interaction – A theory of problematic internet use and psychosocial well-being. *Communication Research, 30*(6), 625–648.

Cheunga, C. M. K., & Thadanib, D. R. (2012). The impact of electronic word-of-mouth communication: A literature analysis and integrative model. *Decision Support Systems, 54*(1), 461–470.

Childers, T. L., & Kaufman-Scarborough, C. (2009). Expanding opportunities for online shoppers with disabilities. *Journal of Business Research, 62*, 572–578.

Choi, J., & Lee, K. (2003). Risk perception and e-shopping: A cross-cultural study. *Journal of Fashion Marketing & Management, 7*(1), 49–64.

Donovan, R. J., & Rossiter, J. R. (1982). Store atmosphere: An environmental psychology approach. *Journal of Retailing, 58*, 34–57.

Donovan, R. J., Rossiter, J. R., Marcoolyn, G., & Nesdale, A. (1994). Store atmosphere and purchasing behavior. *Journal of Retailing, 70*(3), 284–294.

Ebbinghaus, H. (1913). *On memory: A contribution to experimental psychology.* New York: Teachers College.

Greenfield, D. N. (1999). Psychological characteristics of compulsive Internet use: A preliminary analysis. *Cyberpsychology & Behavior, 2*(5), 403–412.

Grewal, D., Roggeveen, A. L., Puccinelli, N. M., & Spence, C. (2014). Retail atmospherics and in-store non-verbal cues: An introduction. *Psychology & Marketing, 31*(7), 469–471.

Joinson, A. N., Reips, U.-D., Buchanan, T., & Paine Schofield, C. B. (2010). Privacy, trust and self-disclosure online. *Human-Computer Interaction, 25*, 1–24.

Kim, J. H., Kim, M., & Lennon, S. J. (2009). Effect of website atmospherics on consumer responses: Music and product presentation. *Direct Marketing and International Journal, 3*(1), 4–19.

La Rose, R., & Eastin, M. S. (2002). Is online buying out of control? Electronic commerce and consumer self-regulation. *Journal of Broadcasting and Electronic Media, 46*(4), 549–564.

Law, D., Wong, C., & Yip, J. (2012). How does visual merchandising affect consumer affective response?: An intimate apparel experience. *European Journal of Marketing, 46*(1/2), 112–133.

Lee, S., & Cude, B. J. (2012). Consumer complaint channel choice in online and offline purchases. *International Journal of Consumer Studies, 36*(1), 90–96.

Lieber, E., & Syverton, C. (2012). Online versus offline competition. In M. Peitz & J. Waldfogel (Eds.), *The Oxford handbook of the digital economy* (pp. 189–223). Oxford University Press.

MacLeod, C. M. (1975). Long-term recognition and recall following directed forgetting. *Journal of Experimental Psychology: Human Learning & Memory, 1*, 271–279.

McGeoch, J. A. (1936). Studies in retroactive inhibition: VII. Retroactive inhibition as a function of the length and frequency of presentation of the interpolated lists. *Journal of Experimental Psychology, 19*, 674–693.

McGeoch, J. A. (1942). *The psychology of human learning.* New York: Longmans, Green.

McKnight, D. H., Choudhury, V., & Kacmar, C. (2002). The impact of initial consumer trust on intentions to transact with a web site: A trust building model. *Journal of Strategic Information Systems, 11*(3–4), 297–323.

Office for National Statistics. (2014). Retrieved May 4, 2015 from http://www.ons.gov.uk/ons/rel/rdit2/internet-access---households-and-individuals/2014/stb-ia-2014.html#tab-Internet-Shopping

Oliver R. L. (1977). Effect of expectation and disconfirmation on postexposure product evaluations – An alternative interpretation. *Journal of Applied Psychology, 62*(4), 480.

Oliver, R. L. (1980). A cognitive model of the antecedents and consequences of satisfaction decisions. *Journal of Marketing Research, 17,* 460–469.

Park, J. H., Lennon, S. J., & Stoel, L. (2005). Online product presentation: Effects on mood, perceived risk, and purchase intention. *Psychology & Marketing, 22*(9), 695–719.

Passyn, K. A., Diriker, M., & Settle, R. B. (2011). Images of online versus shopping: Have the attitudes of men and women, young and old really changed? *Journal of Business & Economics Research, 9*(1), 100–110.

Puccinelli, N. M., Goodstein, R. C., Grewal, D., Price, R., Raghubir, P., & Stewart, D. (2009). Customer experience management in retailing: Understanding the buying process. *Journal of Retailing, 85,* 15–30.

Rose, S., & Dhandayudham, A. (2014). Towards an understanding of internet-based problem shopping behaviour: The concept of online shopping addiction and its proposed predictors. *Journal of Behavioural Addiction, 3*(2), 83–89.

Shankar, V., Smith, A. K., & Rangaswamy, A. (2003). Customer satisfaction and loyalty in online and offline environments. *International Journal of Research in Marketing, 20*(2), 153–175.

Skinner, B. F. (1938). *The behavior of organisms: An experimental analysis.* New York: Appleton-Century.

Spence, C., Puccinelli, N. M., Grewal, D., & Roggeveen, A. L. (2014). Store atmospherics: A multisensory perspective. *Psychology & Marketing, 31*(7), 472–488.

Statista (2015). Retrieved May 4, 2015 from http://www.statista.com/statistics/ 320099/online-and-offline-share-of-shopping-channels-in-the-united-king dom-uk/

Suler, J. (2004). The online disinhibition effect. *Cyberpsychology & Behaviour, 7*(3), 321–326.

Thorndike, E. L. (1920). A constant error in psychological ratings. *Journal of Applied Psychology, 4*(1), 25–29.

Tripp, T. M., & Grégoire, Y. (2011.) Retrieved May 6, 2015 from http://sloan review.mit.edu/article/when-unhappy-customers-strike-back-on-the-internet/

Ward, J. C., & Ostram, A. L. (2006). Complaining to the masses: The role of protest framing in customer-created complaint web sites. *Journal of Consumer Research, (33),* 220–230.

Wu, C., Cheng, F., & Yen, D. C. (2008). The atmospheric factors of online storefront environment design: An empirical experiment in Taiwan. *Information & Management, (45),* 493–498.

Zajonc, R. B. (1968). Attitudinal effects of mere exposure. *Journal of Personality and Social Psychology, 9*(2, Pt. 2), 1–27.

11
Applying Psychology within Games Development: What Can the Gaming Industry Learn from the Discipline?

Linda Kaye

11.1 Social psychology and digital gaming

The rise of gaming as a social pursuit in which players can typically meet, interact, and play alongside one another calls for a consideration of two issues. Firstly, how the social contexts of gameplay impact on players' experiences, and secondly, on a more practical level, how game developers may utilise evidence of this effect to enhance the positive experiences derived from the activity. This will form the basis for this chapter, which will introduce key evidence from psychological theory and research on gaming, and offer practical suggestions towards future game design. Within this, both the "direct," in-game experiences as well as those social processes which operate outside gameplay will be considered in reference to psychological theory and key evidence. This will be introduced here to give readers an insight into the extent to which psychological understanding can provide some suggestion on the development of game features to promote the social experiences of players within gaming.

11.2 Gaming in a social world

There is a large consensus that playing with others enhances gameplay experiences (Gajadhar et al., 2009; Kaye & Bryce, 2012). This is reflected in a range of psychological studies, which have typically compared differential effects of gameplay within different social scenarios. This includes research comparing effects of gaming tasks with human versus computer-controlled opponents. In particular, it has been found that physiological effects such as heart rate are enhanced when playing with human rather than computer-controlled agents (Lim & Reeves, 2010). Additionally, other studies have found that increasing the social presence of others within gaming is associated with enhanced perceptions of positive mood (Gajadhar et al., 2008; Kaye & Bryce, 2014). Indeed, a recent meta-analysis has revealed consistency in this effect across 32 studies examining a range of different psychological outcomes (Fox et al., 2015). This provides consistent evidence of the impact of aspects of social gaming on players' derived experiences.

As well as the role of other players in gameplay, further contextual differences relate to the *types* of social gameplay being experienced. That is, games can be played competitively or cooperatively, and psychological evidence reveals these distinct forms of gameplay to be experienced differently for players (see Fox et al., 2015). For example, hostility and aggressive behaviour have been found to be enhanced within competitive compared to cooperative gaming tasks (Eastin, 2007), indicating some consideration of the task-type on players' experiences within gameplay. With this in mind, clearly there is reason to assume there to be some implications for game development, on the way in which the design of different task-types may impact upon different gameplay experiences. To advance the practical considerations of these issues, focus is given to specific psychological theories which underpin these effects to indicate the way in which game development may build on these ideas. A review of the relevant psychological theory is provided in the next section.

On a general level, the study of Social Psychology considers the way in which we think, feel, and behave within our social worlds. Within this, a key focus is the interaction between individuals and their contexts to consider how this impacts on factors such as attitudes and behaviour. For example, social psychologists address questions such as *"why do people act more aggressively in groups than on their own?"* and *"what factors determine why we form better relationships with some people over others?."* Within this, a number of key theories have been developed which help explain these social psychological processes. Specifically in relation to social gaming, a number of these theories show promise in underpinning these experiences. In particular, social presence theory (Short et al., 1976) provides an explanation of the effects previously outlined when assessing different levels of social presence of agents within gaming tasks. Similarly, social identity theory can underpin the processes associated with group formation and behaviours within and between social groupings (Brown, 1984) that have key relevance to gaming communities. That is, social identity theory explains the way in which we conceptualise ourselves based on our group memberships. Within this, distinctions between in-groups and out-groups are established in which those members who are perceived to occupy the same social group comprise the in-group, whereas those who do not are categorised as the out-group (Tajfel, 1978, 1979; Tajfel & Turner, 1979). On a practical level, this principle can explain the processes and effects, such as inter-group hostility between online gaming groups (conflict between in- and out-group members), as well as intra-group helping behaviours within online teams. Moreover, other psychological theories can provide an insight into the features of games themselves that may promote particular kinds of player experiences. One theory that contributes in this way is flow theory (Csikszentmihalyi, 1975). This will be detailed in the next section. Taken together, psychological knowledge proves a useful resource in understanding player experiences and offers some indication of key features worthy of investigation within future game development.

11.3 Flow

The seminal work of Csikszentmihalyi (1975) first indicated evidence of the experience of flow in which he observed the intense immersion and enjoyment of groups of artists whilst engaged in their work. His subsequent observations revealed these experiences to extend beyond groups of artists to those engaged in a range of activities, including rock-climbing, dancing and chess (Csikzentmihalyi, 1992). Within this, Csikszentmihalyi suggested the flow experience occurs more easily within structurally designed activities allowing for deep levels of concentration and involvement. Here, he proposed that flow occurs when the individual's abilities or skill levels are at equilibrium with the high demands of the particular task being undertaken (Csikszentmihalyi, 1975; Csikszentmihalyi & Csikszentmihalyi, 1988; Massimini & Carli, 1988). He proposed this balance as key to an individual's perceived enjoyment of the activity. Indeed, psychological research extending from this idea has found that flow-like states are associated with a range of positive experiences, such as high levels of arousal, intrinsic motivation, and positive mood (Csikszentmihalyi, 1975, 1982; Ellis et al., 1994; LeFevre, 1988).

In addition to the condition of appropriate skill-challenge balance, the flow experience is said only to occur if the task itself consists of clear goals and feedback. Other characteristics of flow include focused attention, reduced self-consciousness, sense of control, and distorted sense of time (Csikszentmihalyi, 1975). Here it can be seen that flow is a logical framework to apply to gaming, particularly given its prevalence for players (Poels et al., 2007), and in facilitating enjoyment and positive mood following gameplay (Kaye & Bryce, 2014; Klimmt et al., 2007; Smith, 2007). Additionally, the inherent structure of a majority of games, particularly in relation to objectives and performance feedback, presents another logical justification for applying flow in this way (Johnson & Wiles, 2003; Sweetser & Wyeth, 2005). Indeed, recent evidence suggests a range of key design features influential to enhancing flow experiences in mobile games; in particular these are: player power and control, character configuration, and secure scoring systems (Chou et al., 2014). Specifically in relation to scoring systems, for example, those which misreport scores or permit cheating are not deemed particularly conducive to facilitating flow, given frustrations which may interrupt these experiences. Similarly, functionality which permits players to take control over gameplay to the relevant extent is important, particularly given that sense of control is a key characteristic of the flow experience (Csikszentmihalyi, 1975). Evidence from other types of games has found that game genres which are descriptive by nature are more "flow-inducing" compared to those which are narrative or expository-based (Ghonsooly & Hamedi, 2014).

The original flow model proposed four main flow channels which were said to occur as a result of the varying levels of balance between individual skill level and the challenge of the activity (Csikszentmihalyi, 1975). These are: "boredom" (high skill-low challenge), "apathy" (low skill-low challenge), "anxiety" (low skill-high challenge), and the optimal level of "flow" (high skill-high challenge). Although these channels have been readily applied in a number of different contexts, more recent evidence suggests alternative channels within the context of digital gaming (Takatalo & Häkkinen, 2014). Specifically it has been suggested that boredom, apathy, and anxiety should instead be labeled "relaxation," "impassiveness," and "overwhelm," respectively (Takatalo & Häkkinen, 2014). Understanding the multidimensionality of player experiences of digital games is therefore paramount to informing game design and development. In particular, enhancing flow may be helpful for enhancing game loyalty (Chou et al., 2014). Specifically, this suggests that game design which allows an accumulation of challenge, for example through an appropriate levelling system, which mirrors the increasing skill level of specific players, is a key requisite for effective game design in promoting player flow.

Although much psychological research evidences the existence of flow in gaming, less is understood about how flow may be experienced within social forms of gameplay. Whilst the original flow framework did not account for the social dimensions of flow, later acknowledgement has been paid to the concept of a shared sense of flow, interchangeably termed "shared flow," "group flow," or "networked flow" (Gaggioli et al., 2011, 2012; Nakamura & Csikszentmihalyi, 2002; Sawyer, 2008; Sato, 1988; Walker, 2010). For example, evidence from research on motorcycle groups suggests that social dimensions such as companionship and sense of social belonging enhance the "autotelic," rewarding experience originally outlined by Csikszentmihalyi (1975). In this context, "autotelic" refers to the extent to which an individual partakes in an activity purely for the enjoyment and intrinsic value the activity provides, with no external incentive being required to promote such engagement. In relation to social forms of flow, recent research has shown that experiences of flow are higher in more interactive social contexts compared to more coactive ones and related to a sense of group identity (Rufi et al., in press). In a similar way, this has been found to be the case for social gaming, supporting the application of flow in digital gaming contexts (Argenton et al., 2014; Chiang et al., 2011). That is, group flow in cooperative gameplay (i.e. two or more players undertaking a task which requires complementary participation to accomplish a shared goal) occurs as the result of common focus on parallel and organised tasks, which is characterised by indicators such as a shared social belonging and collective competency (Kaye & Bryce, 2012). Indicators such as players' awareness of the task-relevant skills in others have also been found to be a key feature of group flow in competitive forms of gameplay (Kaye & Bryce, 2012). Furthermore, more immersive experiences have been found to be related to features such as interpersonal interactions (Huang et al., 2011). Interestingly, experiences of group flow have been found to represent equivalent flow states to those derived in solo gaming (Kaye & Bryce, 2014). In a similar vein, the concept of social presence can underpin shared social gaming experiences, in which a sense of immersion occurs through group-based tasks (de Kort et al., 2007). Social presence is defined as being an individual's sense of awareness of the presence of another individual through any form of social medium (Short et al., 1976). This can be acknowledged both within face-to-face forms of communication, as well as those experienced through virtual means (e.g., online or LAN gaming). For example, in online instructional contexts, research has identified a number of facets within social context (e.g., familiarity of recipients, trust), online communication (e.g., real-time, language), and interactivity (e.g., types of online task, size of groups) that determine users' perception of social presence (Tu & McIssac, 2002). Within the context of digital gaming, four main facets have been found which are related to a sense of social presence (Hudson & Cairns, 2014). That is, within competitive-based play, "awareness" and "engagement" function as key indicators of social presence, whilst "team involvement" and "perceived team cohesion" operate within cooperative-based gameplay (Hudson & Cairns, 2014).

11.4   Creating "flowing" games

When considering the relevance of these findings for game design and development, there are a number of key implications. That is, in relation to the notions of "collective competency" or "perceived team cohesion," game features which foster collaborative play, in which players are appropriately scaffolded to reach a mutual level of competency and cohesion, and to gain game and player-based feedback, may facilitate such experiences of group flow or social presence, as previously described. Additionally, indicators of group members' competency (e.g., in-game displays of expertise level or accumulative points as measures of performance) are typical features of many games which may foster group-based understanding of collective competence, to facilitate immersive experiences in collaborative gaming. Similarly, such features may also be relevant within competitive-based gaming, in the form of displays of players' skill level relative to others, to foster an awareness of skill level relative to others. Some examples of games which readily adopt this technique include "Trials Frontier" and "Battle Run," in which players are matched with other players of similar competency levels (e.g., based on achievement or player points) to develop effective, competitive forms of gameplay. In relation to "shared sense of social belonging" and likewise, "team involvement," it may be useful to provide identification of an individual player's role and achievements in relation to common group goals or objectives. This may promote a sense of purpose and role within a broader social network. Indeed, the underpinning role of *identification* within group-based game play appears to hold the key to fostering the aforementioned indicators of immersive experiences. That is, on condition that an individual player can *identify* their role in the group, it is only then that a sense of social belonging may be fostered, and in turn, promote a shared sense of flow. Conversely, should an individual player fail to identify their competency relative to others, for example, it may be speculated that this may reduce the extent to which group flow can occur in competitive-based gameplay. Therefore, game development which focuses its attention on understanding the identification process of players in these contexts may be fundamental to such positive, group-based experiences. In this way, acknowledging the wider social contexts and psychological processes of gaming (e.g., group identification) are fundamental to better understanding their impact on these "direct" social experiences that the activity can provide.

11.5 Wider social contexts of gaming

Of course, not all social experiences associated with digital gaming are "direct" in-game experiences with others. In fact, more recently, psychological research has been invested in exploring the concept of "alone together" (Ducheneaut et al., 2006); an experience in which players engage in online forms of gaming (e.g., often massively multiplayer online role-playing games) but do not necessarily engage in what might be considered "social" activities within this context (e.g., social interactions, conversations, team-work). Therefore, although the context of such play is social by nature, the actual form of engagement can be relatively solitary, whereby it is simply the presence of others which forms the basis for their attraction (Ducheneaut et al., 2006). Indeed, this is supported by other research showing how players of Massively Multiplayer Online Role-Playing Games (MMORPGs) use other players in this form of game as a "backdrop" in this context (Williams et al., 2006). That is, although direct socialisation may not be initiated or experienced, the mere social presence of others promotes a sense of social engagement. Given this evidence, consideration is required on the role of wider social experiences of the activity. That is, although direct socialisation experiences may not always be important within online forms of gameplay, opportunities for engagement in gaming communities and social networks may, for example, still be relevant within this context. This assertion is supported by Koster (2005) who maintains that MMORPGs are communities rather than games. Here, the implications for game design relate to providing effective platforms and space for "audiences" in gameplay in which the social presence of others is maintained, yet play may be undertaken independently. Again, psychological insight into the processes such as role identification and player identity within these contexts is potentially useful in understanding the impact on positive player experiences. One useful theoretical framework which may help explain these processes is social identity theory (Tajfel, 1978, 1979; Tajfel & Turner, 1979), particularly in understanding the identities individuals may foster within group-based contexts.

11.6   Creating effective gaming communities

In addition to the "in-game" experiences that socially-based gaming can provide, there is also the potential for games to foster wider networks and community-based experiences for players. Indeed, gaming as a shared activity can function in the formation and development of friendships, promote social identity to particular gaming groups (Kaye, 2014; O'Connor et al., 2015), and can be related to positive psychosocial outcomes, such as psychological well-being (Kaye, 2015; Kaye & Pennington, 2015). This is underpinned by core social psychological principles of social identity theory (Tajfel, 1978, 1979; Tajfel & Turner, 1979). This theory asserts that an individual's sense of self is determined by their degree of affiliation to social groups. Within this, an individual gains a sense of both "personal" and "collective" self, which plays a role in their perceptions of "in-group" and "out-group" members. In particular, those individuals with a high degree of affiliation, as an in-group member, to any given group are said to experience a high sense of self-esteem through such affiliation (Hogg & Abrams, 1990). Additionally, these positive appraisals translate into wider positive outcomes for feelings of life satisfaction and well-being (Isiklar, 2012; Kong et al., 2013). It is therefore important to consider how these principles may be applied within the design of games and gaming communities. In particular, this may be achieved by considering the three inter-related processes which underpin social identity theory (Tajfel, 1978). These are:

1. *Social categorisation*, in which individuals see themselves and others as categories rather than as individuals.
2. *Social comparison*, in which individuals assess the worth of groups through comparing their relative features.
3. *Social identification*, in which individuals' identities are formulated by their experiences within a social group.

Some evidence is available to suggest the practical applications of these ideas. Namely, Jones (1998) proposes the role of discourses within online communities which help form connections and community spirit between individuals. This may consist of considering whether there is a culturally-based distinction within specific gaming groups compared to others. Within this, identification or development of specific "group language" could provide a useful way of fostering these processes. For example, acronyms that are readily used by MMORPG players are one of the features through which this can be fostered. Additionally, when considering *social comparison*, some explicit indication of the values and benefits of group membership to specific gaming communities may prompt *evaluations of worth* for "in-group" membership, compared to membership of other groups.

Additionally, the notion of promoting "in-group" membership through these forms of design also has relevance for effective integration of players within the gaming context. This is particularly important as a means of reducing the likelihood of any negative behaviours as a result of prejudices amongst sub-groups. For example, it is commonly reported that female online gamers are subject to discrimination within gameplay and are often perceived as passive or dependent on men (Mou & Peng, 2008). This is further supported by research showing that male characters engage more frequently in leadership and achievement-based tasks compared to females (Thompson & Zerbinos, 1995). These gender-based roles are reinforced by the typical representation of female characters in digital games. Indeed, it has been found that a large percentage do not include any female characters at all (Williams et al., 2009), and in those that do, female characters are often given secondary roles and portrayed in overly sexualised ways (Provenzo, 2000; Williams et al., 2009). These representations may therefore promote negative gender-based stereotypes relating to females in gaming and influence negative behaviours. This notion is supported by evidence highlighting that negative stereotypes do indeed influence "real world" prejudices (Brenick et al., 2007; Cicchirllo 2009; Dill & Burgess, 2012), as well as potentially having negative effects on gameplay performance in female gamers (Kaye & Pennington, 2015). These effects are underpinned by expectancy theory (Berger et al., 1977) suggesting that cultural norms determine expected behaviour. In relation to gender, this highlights how men and women are expected to behave, which subsequently shapes the responses of others to any named behaviour (Berger et al., 1977). For example, men are typically assumed to be more dominant than women, who in contrast are seen as more submissive. In the case of online gaming, this is indeed consistent, particularly considering that online gaming is typically characterised as a male-dominated environment, where men are considered higher status than women (Ivory et al., 2014). Non-compliance with these assumed behaviours results in negative social responses (Fox & Tang, 2014; Slater & Blodgett, 2012). Particularly in relation to performance outcomes, the negative effect of stereotypes has been well documented. This has been referred to as "stereotype threat" (Steele & Aronson, 1995), and refers to circumstances in which an individual's performance is impeded by stereotype-salient cues. This effect has been typically found within intelligence tests, memory tasks, and other practical-based tasks in which an imposed threat reduces performance (e.g., Skorich et al., 2013; Steele & Aronson, 1995). However, the implications for female-based stereotype threats to gameplay performance and behaviours are highly pertinent (Kaye & Pennington, 2015). For example, in team-based gaming, discriminatory behaviour (e.g., over text or voice-based communication) towards female players may impact negatively on performance in meeting the demands of the game and team objectives. The utility of digital games and communities as platforms for alleviating gender-based roles (through modifications which are not readily accessible in "real world" contexts) may hold implications for reducing gender divides in contexts beyond the gaming world.

Understanding the social processes which underpin stereotypes as a means of reducing their impacts is clearly an important area of consideration. The processes of social identity theory can provide some means of fostering more effective social dynamics within group membership, to alleviate the likelihood of gender-based stereotypes (or indeed, other observed stereotypes within gaming contexts) impeding in-game behaviours and performance. Within this, one practical consideration is the way in which game environments and communities may promote a broader sense of in-group membership, through the principles of adopting multiple social identities ("I am an online gamer," rather than "I am a *female* online gamer"). Here, players may experience affiliation to an alternative social identity in an attempt to realise the observed benefits to self-esteem (Rydell & Boucher, 2010; Rydell et al., 2009), and reduce the negative effects of stereotypes, such as those relating to gender distinctions. Indeed, research indicates the role of "multiple identities" in widening individuals' self-categorisations of group membership (Rydell et al., 2009). It has been suggested that when their social identity is threatened, individuals categorise themselves with an alternative positive social identity as a means of maintaining positive self-esteem (Rydell & Boucher, 2010; Rydell et al., 2009). Given that some forms of digital gaming (and indeed other virtual platforms) allow individuals to explore alternative forms of identity to those they occupy in the "real world" (discussed in the next section), affiliating to these alternative identities can present opportunities to maintain self-esteem against other instances which may threaten it. In particular, platforms which aims to foster social belonging and the integration of individuals might best facilitate this process, particularly for those who may experience stigmatisation of an aspect of their identity (e.g, gender, race, ethnicity, disability) in the "real world." In the case of stereotyped gaming groups, those who may feel stigmatised based on one identity (e.g., as a female gamer) may seek alternative positive identities (e.g., as an experienced online gamer) to alleviate negative personal effects. In addition to reducing negative ramifications for the individual, this widening identity strategy may also be used to remove gendered labelling across a gaming community, to promote a sense of "sameness" across a range of players, regardless of gender (or other distinguishing features which may promote narrow in-group versus out-group categorisations). Likewise, these principles may be applied in a wider context by considering that reduced group categorisation in gaming could transfer beyond this context, particularly for those individuals who engage with the same people in gaming communities and in other "real world" contexts. That is, reducing stigmatisation in gaming contexts, through strategies such as removing labelling, and acknowledging wider identities of individuals, may transfer into more favourable real-world attitudes towards stigmatised individuals and groups.

11.7 Creating a "sense of self"

Furthering the notion of social identities, another interesting psychological aspect of gaming relates to presentation of identity or self in digital games. That is, a large majority of games allow some form of physical representation of self through user-generated avatars. Although there is great diversity across games in the flexibility of these representations, evidence points to their psychological function for players in a range of different ways. One such psychological function relates to identity formation and exploration, which has been observed across a range of studies, in different gaming contexts (e.g., Konijn & Bijvank, 2009; Ritterfield, 2009; Turkle, 1994). For example, evidence suggests that MMORPG players create "ideal selves" which consist of more favourable attributes than their own, suggesting this platform as an effective means of identity exploration for these players (Bessière et al., 2007). Similarly, in other virtual environments, such as Second Life, the platform permits users to present themselves physically as their desired identity (Kleban & Kaye, 2015; Neustaedter & Fedorovskaya, 2009). Additionally, avatar physicality in particular has been found to have impacts on both in-game behaviours, such as player performance, and subsequent real world attitudes and interpersonal behaviours (Yee & Bailenson, 2006, 2007; Yee et al., 2009). For example, evidence for the so-called "Proteus effect" has found that physical characteristics of avatars (e.g., attractiveness, height) affect the way in which individuals who occupy these will subsequently interact with people (e.g., friendliness to others, aggressiveness to others) (Yee & Bailenson, 2007). Finally, an interesting point to note is that the extent to which players explore alternative identities is found to relate to personality dimensions (Yee, 2009). That is, more extrovert individuals (i.e. those who are highly sociable) are more likely to experiment with identity exploration compared to those who are less extrovert (Yee, 2009). Clearly, understanding the way in which avatars and other indicators of player identity can function for individuals is an area of practical significance for games development, as well as other virtual platforms. Indeed, scholars have identified the notion of "second selves" in the way in which Internet users develop online representations of themselves (Kafai et al., 2007; Turkle, 1984), highlighting the need for digital spaces to promote these opportunities. In particular, providing a range of customisation options for identity portrayal in digital games is one key recommendation for facilitating players' identity exploration. This may allow opportunities for "identity tourism" (Taylor, 2002), in which individuals are able to experiment with identities which reflect distinctions from themselves, such as gender swapping (Hussain & Griffiths, 2008), which are not so readily achievable beyond the virtual world. Additionally, previous research findings have revealed that identity in MMORPGs results in immersive experiences. (Nojima, 2007) calls for a greater understanding of such identity processes in digital games. In particular, it is recommended that game designers consider the extent to which their games allow the following identity portrayals:

- The "actual" self – a representation of an individual's typical conception of him/herself
- The "ideal" self – a representation of how an individual would wish to be in an ideal world
- The "ought" self – a representation of how an individual conceives they should be, often based on societal values

These are based on the work of Higgins (1987), in which he suggested that negative emotionality arises when disparity is recognised between an individual's conception of their actual and ideal selves. Indeed, this has been found to be somewhat consistent in MMORPG players, amongst whom those with lower psychological well-being attribute more favourable features to their character than themselves (Bessière et al., 2007). Similarly, other research has found that self-discrepancies between virtual and real selves are negatively related to psychological states of autonomy and recovery (Suh, 2013), suggesting the principles of Higgins' (1987) theory play out in somewhat equivalent ways within some contemporary contexts. However, these effects have not yet been established in line with a number of key issues. Firstly, these principles may not apply to all virtual contexts, including all types of digital game. That is, some research indicates that individuals with physical disabilities experience heightened self-esteem from representing themselves without their disability in virtual contexts (Kleban & Kaye, 2015), suggesting that discrepancies between ideal and actual selves in virtual contexts are not always associated with negative emotionality. This in itself highlights the importance of allowing opportunities for alternative identity portrayals in virtual contexts, such as avatar choice and/or customisation options which represent a variety of identities that may be alternatives from "real world ones." Secondly, understanding users' motivations for different virtual identity portrayals requires further insight to help industry representatives understand the range of reasons underpinning avatar choice and customisation and their association with users' psychological and emotional experiences. The role of virtual identities should therefore be considered a key issue within games design and development given its role in players' psychological experiences. In particular, key recommendations might be:

- Providing opportunities for players to utilise avatars/characters representing a range of genders and sexualities, which do not necessarily determine the direction of gameplay structure, format, or storyline, for example.
- Providing a range of customisation options for players to explore a range of physical identity portrayals (e.g., size, weight, race).
- Providing players with opportunities to provide additional written details to supplement their virtual identity portrayals, to permit further disclosures of identity.

11.8   Conclusion

The social experiences that gaming can provide highlight the utility of applying established social psychological understanding to help establish the processes underpinning positive experiences for players. These relate both to in-game experiences, such as flow, but also the wider social narratives present within gaming communities and their interactions. Fostering effective social integration is particularly key to enhancing positive experiences for players. This is relevant both within and beyond gameplay. Indeed, this chapter highlights the role of player identification within a group as one key process that underpins immersive group gameplay, and its wider application in effective interpersonal relations between players beyond gameplay (as underpinned by social identity theory). Therefore, investment in further understanding of these identification processes is a key recommendation to aid effective game design and development. Although the theoretical underpinnings are useful, these require greater practical enquiry to better appreciate their relevance in practice.

11.9   References

Argenton, L., Triberti, S., Serino, S., Muzio, M., & Riva, G. (2014). Serious games as positive technologies for individual and group flourishing. In A. L. Brooks, S. Brahman, & L. C. Jain (Eds.), *Technologies of inclusive well-being* (pp. 221–244). Berlin and Heidelberg: Springer.

Berger, J., Fisek, H., Norman, R., & Zelditch, M. (1977). *Status characteristics and social interaction.* New York: Elsevier.

Bessière, K., Seay, A. F., & Kiesler, S. (2007). The ideal Elf: Identity exploration in World of Warcraft. *Cyberpsychology & Behavior, 10*(4), 530–535.

Brenick, A., Henning, A., Killen, A., O'Connor, M., & Collins, M. (2007). Social evaluations of stereotypic images in video games: Unfair, legitimate, or "just entertainment"? *Youth and Society, 38*(4), 395–419.

Brown, R. J. (1984). The effects of intergroup similarity and cooperative versus competitive orientation on intergroup discrimination. *British Journal of Social Psychology, 23*(1), 21–33.

Chiang, Y. T., Lin, S. S., Cheng, C. Y., & Liu, E. Z. F. (2011). Exploring online game players' flow experiences and positive affect. *Turkish Online Journal of Educational Technology, 10*(1), 106–114.

Chou, J. C., Hung, C., & Hung, Y. (2014, September). *Design factors of mobile game for increasing gamers' flow experience.* IEEE International Conference on Management of Innovation and Technology 2014 (pp. 137–139). Singapore.

Cicchirillo, V. J. (2009). *The effects of priming racial stereotypes through violent video games.* Dissertation Abstracts International Section A: Humanities and Social Sciences, 3683–3683. Retrieved July 10, 2014 from https://etd.ohiolink.edu/

Csikszentmihalyi, M. (1975). *Beyond boredom and anxiety: Experiencing flow in work and play.* San Francisco: Jossey-Bass Publishers.

Csikszentmihalyi, M. (1982). Towards a psychology of optimal experience. In L. Wheeler (Ed.), *Review of personality and social psychology* (Vol. 3, pp. 13–36). Beverly Hills, CA: Sage.

Csikszentmihalyi, M. (1992). *Flow: The psychology of happiness.* London: Rider.

Csikszentmihalyi, M., & Csikszentmihalyi, I. S. (1988). *Optimal experience: Psychological studies of flow in consciousness.* New York: Cambridge University Press.

de Kort, Y. A. W., IJsselsteijn, W. A., & Poels, K. (2007, October). *Digital games as social presence technology: Development of the social presence in gaming questionnaire (SPGQ).* Paper Presented at PRESENCE Conference 2007, Barcelona, Spain.

Dill, K. E., & Burgess, M. C. R. (2012). Influence of black masculinity game exemplars on social judgments. *Simulation & Gaming, 44* (1), 1–24

Ducheneaut, N., Yee, N., Nickell, E., & Moore, R. J. (2006, April). *'Alone together?' exploring the social dynamics of massively multiplayer online games.* ACM Conference on Human Factors in Computing Systems (Montreal; Canada. NY).

Eastin, M. S. (2007). The influence if competitive and cooperative group game play on state hostility. *Human Communication Research, 33*(4), 450–466.

Ellis, G. D., Voelkl, J. E., & Morris, C. (1994). Measurement and analysis issues with explanation of variance in daily experience using the flow model. *Journal of Leisure Research, 26*(4), 337–356.

Fox, J., Ahn, S. J., Janssen, J. H.,Yeykelis, L., Segovia, K. Y., & Bailenson, J. N. (2015). Avatars versus agents: A meta-analysis quantifying the effect of agency on social influence. *Human Computer Interaction, 30*(5), 401–432.

Fox, J., & Tang, W. Y. (2014). Sexism in online video games: The role of conformity to masculine norms and social dominace orientation. *Computers in Human Behavior, 33,* 314–320.

Gaggioli, A., Milani, L., Mazzoni, E., & Riva, G. (2011). Networked flow: A framework for understanding the dynamics of creative collaboration in educational and training settings. *The Open Education Journal, 4*(1), 41–49.

Gaggioli, A., Riva, G., Milani, L., & Mazzoni, E. (2012*). Networked flow: Towards an understanding of creative networks.* Dordrecht: Springer Science & Business Media.

Gajadhar, B. J., de Kort, Y. A., & Ijsselsteijn, W. A. (2008, April). *Influence of social setting on player experience of digital games.* Paper Presented at CHI 2008 Conference, Florence, Italy.

Gajadhar, B. J., de Kort, Y. A. W., IJsselsteijn, W. A., & Poels, K. (2009, October). *Where everybody knows your game: The appeal and function of game cafes in Western Europe.* Paper Presented at the International Conference on Advances in Computer Entertainment Technology, Athens, Greece.

Ghonsooly, B., & Hamedi, S. M. (2014). An investigation of the most flow inducing genres. *International Journal of Research Studies in Education, 3*(4), 1–10

Higgins, E. T. (1987). Self-discrepancy: A theory relating self and affect. *Psychological Review, 94*(3), 319–340.

Hogg, M. A., & Abrams, D. (1990). Social motivation, self-esteem and social identity. In D. Abrams & M. A. Hogg (Eds.), *Social identity theory: Constructive and critical advances* (pp. 28–47). London: Harvester Wheatsheaf.

Huang, L.-T., Chiu, C.-A., Sung, K., Farn, C.-K. (2011). A comparative study on the flow experience in web-based and text-based interaction environments. *Cyberpsychology, Behavior, and Social Networking, 14*(1–2), 3–11.

Hudson, M., & Cairns, P. (2014). Measuring social presence in team-based digital games. In G. Riva, J. Waterworth, & D. Murray (Eds.*), Interacting with presence: HCI and the sense of presence in computer-mediated environments* (pp. 83–101). Warsaw: De Gruyter Open.

Hussain, Z., & Griffiths, M. D. (2008). Gender swapping and socializing in cyberspace: An exploratory study. *CyberPsychology & Behavior, 11*(1), 47–53.

Isiklar, A., 2012. Examining psychological well being and self esteem levels of Turkish students in gaining identity against role during conflict periods. *Journal of Instructional Psychology, 39*(1), 41–50.

Ivory, A. H., Fox, J., Waddell, F., & Ivory, J. D. (2014). Sex role stereotyping is hard to kill: A field experiment measuring social responses to user characteristics and behaviour in an online multiplayer first-person shooter game. *Computers in Human Behavior, 35,* 148–156

Johnson, D., & Wiles, J. (2003). Effective affective user interface design in games. *Ergonomics, 46*(13/14), 1332–1345.

Jones, S. G. (1998). *Cybersociety 2.0: Revisting computer-mediated communication and community.* London: Sage Publications Inc.

Kafai, Y. B., Fields, D. A., & Cook, M. (2007). *Your second selves: Avatar designs and identity play in a teen virtual world.* Proceedings of the Situated Play: DiGRA Conference (Tokyo, Japan).

Kaye, L. K. (2014). Football manager as a persuasive game for social identity formation. In D. Ruggiero (Ed.), *Cases on societal effects of persuasive games* (pp. 1–17). USA: IGI Global.

Kaye, L. K. (2015). *Social identity as a predictor of self-esteem and psychological well-being in a sample of digital gamers.* British Psychological Society Annual Conference 2015 (ACC Liverpool, UK).

Kaye, L. K., & Bryce, J. (2012). Putting the 'fun factor' into gaming: The influence of social contexts on experiences of playing videogames. *International Journal of Internet Science, 7*(1), 23–37.

Kaye, L. K., & Bryce, J. (2014). Go with the flow: The experience and affective outcomes of solo versus social gameplay. *Journal of Gaming and Virtual Worlds, 6*(1), 49–60.

Kaye., L. K., & Pennington, C. (2015). *Applications of social identity theory in the context of digital gaming.* Department of Psychology Research Seminar Series 2014–2015 (Edge Hill University, UK).

Kleban, C., & Kaye, L. K. (2015). Psychosocial impacts of engaging in Second Life for individuals with physical disabilities. *Computers in Human Behavior, 45*, 59–68.

Klimmt, C., Hartmann, T., & Frey, A. (2007). Effectance and control as determinants of video game enjoyment. *CyberPsychology & Behavior, 10*(6), 845–848.

Kong, F., Zhao, J., & You, X. (2013). Self-esteem as mediator and moderator of relationship between social support and subjective well-being among Chinese university students. *Social Indicators Research, 112*(1), 151–161.

Konijn, E. A., & Bijvank, M. N. (2009). Doors to another me: Identity construction through digital gameplay. In U. Ritterfield, M. Cody, & P. Vorderer (Eds.), *Serious games: Mechanics and effects* (pp. 179–203). Oxon: Taylor and Francis.

Koster, R. (2005). Gaming. Retrieved January 9, 2015 from http://www.raphkoster.com/gaming/

LeFevre, J. (1988). Flow and the quality of experience during work and leisure. In. M Csikszentmihalyi & I. S. Csikszentmihalyi (Eds.), *Optimal experience: Psychological studies of flow in consciousness* (pp. 307–318). New York: Cambridge University Press.

Lim, S., & Reeves, B. (2010). Computer agents versus avatars: Responses to interactive game characters controlled by a computer or other player. *International Journal of Human-Computer Studies, 68* (1–2), 57–68.

Massimini, F., & Carli, M. (1988). The systematic assessment of flow in daily experience. In M. Csikszentmihalyi & I. S. Csikszentmihalyi (Eds.), *Optimal experience: Psychological studies of flow in consciousness* (pp. 266–287). Cambridge: Cambridge University Press.

Mou, Y., & Peng, W. (2008). Gender and racial Stereotypes in popular video games. In. R. Ferdig (Ed.), *Handbook of Research on Effective Electronic Gaming in Education* (pp. 922–937). Hershey, PA: IGI Global.

Nakamura, J., & Csikszentmihalyi, M. (2002). The concept of flow. In C. R. Snyder & S. J. Lopez (Eds.), *Handbook of positive psychology* (pp. 89–105). Oxford: Oxford University Press.

Neustaedter, C., & Fedorovskaya, E. (2009, May). *Presenting identity in a virtual world through avatar appearance*. Proceedings of Graphics Interface. (Toronto, Canada).

Nojima, M. (2007, September). *Pricing models and motivations for MMO play*. Proceedings of DiGRA 2007 (Tokyo, Japan).

O'Connor, E. L., Longman, H., White, K. M., & Obst, P. L. (2015). Sense of community, social identity and social support among players of Massively multiplayer online games (MMOGs): A qualitative analysis. *Journal of Community and Applied Social Psychology, 25*(6), 459–473. doi: 10.1002/casp.2224

Poels, K., de Kort, Y. A. W., & IJsselsteijn, W. A. (2007, November). *'It is always a lot of fun!' Exploring dimensions of digital game experience using focus group methodology*. Paper Presented at the Futureplay 2007, Toronto, Canada.

Provenzo, E.F. (2000). Computing, Culture, and Educational Studies. Educational Studies: *Journal of the American Educational Studies Association, 31*(1), 5–19.

Ritterfield, U. (2009). Identity formation and emotion regulation in digital gaming. In U. Ritterfield, M. Cody, & P. Vorderer (Eds.), *Serious games: Mechanics and effects* (pp. 204–218). Oxon: Taylor and Francis.

Rufi, S., Wlodarczyk, A., Páez, D., & Javaloy, F. (in press). Flow and emotional experience in spirituality: Differences in interactive and coactive collective rituals. *Journal of Humanistic Psychology*

Rydell, R. J., & Boucher, K. L. (2010). Capitalizing on multiple social identities to prevent stereotype threat: The moderating role of self-esteem. *Personality and Social Psychology Bulletin, 36*(2), 239–250.

Rydell, R. J., McConnell, A. R., & Beilock, S. L. (2009). Multiple social identities and stereotype threat: Imbalance, accessibility, and working memory. *Journal of Personality and Social Psychology, 96*(5), 949–966.

Sato, I. (1988). Bosozoku: Flow in Japanese motorcycle gangs. In M. Csikszentmihalyi & I. Csikszentmihalyi (Eds.), *Optimal experience: Psychological studies of flow in consciousness* (pp. 92–117). Cambridge: Cambridge University Press.

Sawyer, K. (2008). *Group genius: The creative power of collaboration*. New York: Basic Books.

Short, J., Williams, E., & Christie, B. (1976). *The social psychology of telecommunications*. London: John Wiley & Sons, Ltd.

Skorich, D. P., Webb, H., Stewart, L., Kostyanaya, M., Cruwyz, T., McNeill, K., & O'Brien, K. J. (2013). Stereotype threat and hazard perception among provisional license drivers. *Accident Analysis and Prevention, 54*, 39–45.

Salter, A. & Blodgett, B. (2012) Hypermasculinity & Dickwolves: The Invisibility of Women in the New Gaming Public. *Journal of Broadcasting & Electronic Media. 56*(3), 401–416.

Smith, B. P. (2007). *Flow and the enjoyment of video games*. (Unpublished doctoral dissertation). University of Alabama, USA.

Steele, C. M., & Aronson, J. (1995). Stereotype threat and the intellectual test performance of African Americans. *Journal of Personality and Social Psychology, 69*(5), 797–811.

Suh, A. (2013). The influence of self-discrepancy between the virtual and real selves in virtual communities. *Computers in Human Behavior, 29*(1), 246–256.

Sweetser, P., & Wyeth, P. (2005). GameFlow: A model for evaluating player enjoyment in games. *ACM Computers in Entertainment, 3*(3), 1–24.

Tajfel, H. (1978). *Differentiation between social groups*. London: Academic Press.

Tajfel, H. (1979). Individuals and groups in social psychology. *British Journal of Social and Clinical Psychology, 18*(2), 183–190.

Tajfel, H., & Turner, J. (1979). An integrative theory of inter-group conflict. In J. A. Williams & S. Worchel (Eds.), *The social psychology of inter-group relations* (pp. 33–47). Belmont, CA: Wadsworth.

Takatalo, J. M. E., & Häkkinen, J. P. (2014). Profiling user experience in digital games with the flow model. In. *Proceedings of the 8th Nordic conference on human-computer interaction: Fun, fast, foundational* (pp. 353–356). New York.

Taylor, T. L. (2002). Living digitally: Embodiment in virtual worlds. In R. Schroeder (Ed.), *The social life of avatars: Presence and interaction in shared virtual environments* (pp. 40–62). London: Springer-Verlag.

Thompson, T. L., & Zerbinos, E. (1995). Gender roles in animated cartoons: Has the picture changed in 20 years? *Sex Roles, 32,* 651–673.

Tu, C. H., & McIssac, M. (2002). The relationship of social presence and interaction in online classes. *American Journal of Distance Education, 16*(3), 131–150.

Turkle, S. (1984). *The second self: Computer and the human spirit*. New York: Simon & Schuster.

Turkle, S. (1994). Constructions and reconstructions of self in virtual reality: Playing in the MUDs. *Mind, Culture and Activity, 1*(3), 158–167.

Walker, C. J. (2010). Experiencing flow: Is doing it together better than doing it alone? *The Journal of Positive Psychology, 5*(10), 3–11.

Williams, D., Ducheneaut, N., Xiong, L., Zhang, Y., Yee, N., & Nickell, E. (2006). From tree house to barracks: The social life of Guilds in World of Warcraft. *Games and Culture, 1*(4), 338–361.

Williams, D., Martins, N., Consalvo, M., & Ivory, J. D. (2009). The virtual census: Representations of gender race and age in video games. *New Media & Society, 11*(5), 815–834. doi:10.1177/1461444809105354

Yee, N. (2009). Identity projection. *The Daedalus Project: The Psychology of MMORPGs, 7*(1). Retrieved from http://www.nickyee.com/daedalus/archives/000431.php

Yee, N., & Bailenson, J. (2006, August). *Walk a mile in digital shoes: The impact of embodied perspective-taking on the reduction of negative stereotyping in immersive virtual environments*. Proceedings of PRESENCE 2006: The 9th Annual International Workshop on Presence (Cleveland, Ohio, US)

Yee, N., & Bailenson, J. (2007). The Proteus Effect: The effect of transformed self-representation on behaviour. *Human Communication Research, 33*(3), 271–290.

Yee, N., Bailenson, J. N., & Ducheneaut, N. (2009). The Proteus Effect: Implications of transformed digital self-representation on online and offline behaviour. *Communication Research, 36*(2), 285–312.

12
Military and Defence Applications

Coral Dando and Claire Tranter

12.1 Introduction

Virtual environments are synthetic computer simulations that represent activities at a high degree of realism, and which are presented to a user in such a way that s/he temporarily suspends belief and accepts them as real environments (see Witmer & Singer, 1998). Virtual Environments (VEs) allow people to communicate as *avatars*, which are digital visual projections that represent a synthetic reality (Fox & Ahn, 2013), so that individuals can change aspects of their identity, or even create a novel, entirely fictitious, and unrepresentative online identity. Virtual environments have numerous applications for military and defence purposes, ranging from allowing personnel to experience realistic high-pressure situations with a sense of presence but in the absence of real-world risk, to modelling threats to national and international infrastructure to improve resilience. Additional and emerging opportunities also exist for communication and intelligence gathering purposes, exploring online social cognition and group behaviour, and for understanding how to mitigate the negative effects of combat-related stress disorders, for example.

In this chapter we introduce psychological theory and contemporary cyberpsychology research, and offer an albeit very brief introduction to the rapidly developing application of technology for understanding human behaviour and facilitating performance to support and advance military and defence capability. For the purposes of this text we make a distinction between military and defence, and have split the chapter into two distinct sections, accordingly. The nature of defence has changed considerably in the past two decades to encompass the police and intelligence agencies far more than has previously been the case. Broadly speaking, we use the term military to refer to non-civilian armed forces (e.g., Navy, Army & Air Force) that are authorised to use force to support the interests of the state and its citizens. We use the term defence to refer to civilian, non-military law enforcers and investigators who are tasked with proactively and reactively investigating crime and defending national and international infrastructure (e.g., police, security services, and national crime agencies), and who are not typically members of the armed forces. We make this distinction because, while there is clearly commonality, in that both military and defence personnel/organisations are concerned with national (and increasingly international) defence and security (e.g., Defence Reform Act, 2014; Intelligence Services Act, 1994; Ministry of Defence Police Act, 1984), remits and operational environments can differ markedly. Hence, the demands and challenges faced by each are often disparate, and so psychologically-guided cyber research tends to be bespoke, problem-specific, and end-user driven, particularly given that traditional face-to-face (FtF) psychological theories and explanations cannot simply be applied to online behaviours (e.g., Dando & Bull, 2011; Dando et al., 2015; Dando & Tranter, 2015; Kolasinski, 1995; Manojlovich et al., 2003; Taylor et al., 2014).

12.2   Military applications

12.2.1   Training

Perhaps one of the most obvious areas of interest in terms of practical applications of cyberpsychology research is for military training purposes. The military face numerous challenges in meeting the high levels of training necessary to respond effecitvely and efficiently to existing and emerging threats, often at short notice. Operational excellence and mission success rely on personnel becoming combat proficient, and remaining so despite a low base rate of real-life occurences, and limited access to traditional, live, large-scale training environments. Virtual environments offer large numbers of personnel the chance to interact in realistic, simulated face-to-face environments with other distant military units through the Internet (or through the classified network known as SIPRNET), with first-responder units, civilians, and even medical personnel providing a training experience that is increasingly effective, but at a much lower cost than would be required for a real-life training exercise (CRS Report for Congress: Wilson, 2008). Empirical research investigating the effectiveness of large-scale virtual military training is scant because measuring efficacy is challenging, largely due to the fact that defining quantitative outcomes is complex. That said, the military have not developed virtual training in a vacuum; rather, it has exploited positive research findings from other domains, for example, serious gaming technology, which has been adapted by the military for security, healthcare, and communication training (Djaouti et al., 2011).

Using technology to simulate real world action is not new. World War II (WWII) pilots were trained, in part, for example, using flight simulators, which offered opportunities to practice manoeuvres and procedures (Macedonia, 2002). Flight simulator technology has evolved considerably since WWII, and while pilots still rehearse basic procedural tasks without having to worry about actually flying aircraft, current virtual flight simulation technology provides a truly immersive and realistic experience. The practical, procedural benefits remain, but significant additional benefits arise from, among other things, increased awareness and understanding of the psychological impact of "real world" combat. Virtual flight simulators allow pilots (be they military or civilian) to safely and repeatedly experience adverse, life-threatening occurrences. Understanding how human sensory and perceptual systems react (visual, auditory, and other components) and interact in such circumstances allows both pilots and trainers to experience the types of physiological and psychological reactions that can occur under such circumstances, and, importantly, how to manage these reactions to best effect. For example, desensitising pilots to what might otherwise be psychologically and physiologically debilitating occurances when first experienced.

There is considerable merit in allowing military personnel to compare performance in the real world with performance in a virtual environment, particularly if the virtual environment is mimicking the real world. This means that metrics developed for the real world can be deployed in the virtual environment, and vice versa. However, this presumes humans perform similarly in real and virtual environments, which is an important consideration for training applications where a virtual environment is used to train a particular skill that is to be transferred into the real world. Research investigating whether and how skills learned in a VE are transferred to the real world typically reveals equivalent levels of post-training performance (e.g., Lee, 2006; Rose et al., 2000). Virtual- and real-trained real world performance has not been found to result in differences in susceptibility to cognitive and motor interference tasks in terms of spare attention capacity to respond to additional stimuli not directly related to the task demands. In fact, real task performance after training in a VE has been found to be less affected by concurrently performed interference tasks than real task performance after training on the real task, indicating that virtual training can result in equivalent or even better real-world performance than real training (also see Lathan et al., 2002).

The use of virtual environments for training is also being integrated into military medical and surgical training, whereby doctors are able to practice working with limited resources in armed conflict situations before moving from civilian practice to front line field hospitals. This allows learning agendas to be deconstructed and directed toward task-/situation-specific objectives, such as those associated with complex multiple battle injuries, rather than the availability of patients, allowing skill-development to progress prior to real-world military application (Teteris et al., 2012). Some consider that the absence of practice using virtual environments and simulation training to be unethical (Ziv et al., 2003), and for certain high-risk, time-critical medical procedures, simulator training is compulsory (Gallagher & Cates, 2004), in an attempt to protect military patients from unnecessary risk.

The initial introduction of military personnel to a new environment under combat situations is estimated to result in a 40% mortality rate within the first three months of deployment (Eshel, 2011). To reduce the number of fatalities and appropriately equip foot soldiers, immersive simulations for training purposes have been rapidly evolving. Immersive military training simulators initially focused on vehicle operations using physical devices (such as a cockpit or cabin), whereas virtual training for dismounted foot soldiers has only been widely employed in the last decade. This is due, in part, to the increased availability of high quality, affordable interactive technology, comprising highly-defined graphics, often fixed on a head-mounted display (HMD), to create a truly immersive experience (Cruz-Neira et al., 2011). This type of technology allows personnel to prepare better for potentially dangerous situations, because immersive interactions can be made unpredictable and highly realistic. For example, modelling a busy checkpoint in Baghdad, where avatars simulate civilians, portraying facial expressions and emotions, thereby supporting trainees in interpreting and reacting accordingly (here to prevent a roadside bomb from being ignited: Wilson, 2008).

The US Army has recently created the Dismounted Soldier Training System (DSTS; Wang et al., 2012), allowing the simultaneous and interactive training of dismounted squad teams. Each team soldier is equipped with a head-mounted display (HMD), tracking sensor, stereo speakers, microphone headset, 3D display processer, and an instrumented weapon (*see* Figures 12.1 and 12.2). Each trainee controls and is represented by an avatar and all can see and interact with each other

Figure 12.1 US Army soldiers conduct training using the Dismounted Soldier Training System (DSTS) at the 7th Army Joint Multinational Training Command (JMTC) at Grafenwoehr, Germany, 2013

Figure 12.2 Paratroopers conduct simulated missions in 2012, using the DSTS, a virtual reality environment with unlimited mission possibilities

(Intelligent Decisions, 2011). This allows commanders and trainers to deliver collective tasks in challenging terrains, comprising multiple concurrent objectives for maximum skill acquisition. The technology permits soldiers to move and interact within the VE in a natural and ordinary way, enabling them to signal to others and lean/jump around obstacles, ensuring generalisability to real-world combat situations. Military virtual realism is considered much more advanced than civilian comparisons, with enhanced levels of detail, such as complex facial expressions and even disturbed soil, which could indicate a hidden device (Intelligent Decisions, 2011). VEs allow highly detailed levels of interaction, which support effective interpretation of motivations and reactions of other avatars, which can be used to enhance the efficacy of military training. For example, Gehlbach et al. (2012) found that interactive feedback improved the accuracy of social perspective-taking, which in turn improved the accuracy of detecting biases, generating initial hypotheses, and adapting these hypotheses in light of new evidence; all of which can affect the success, or otherwise, of a deployment.

Five primary themes are evidenced in the training – major combat operations, irregular warfare, peace operations, limited intervention and peacetime military engagement – as well as the four elements of Unified Land Operations – offence, defence, stability, and civil support. Inclusive operational themes allow wide-ranging levels of complex motor and cognitive training, maximum skill acquisition and increased validity for real world experiences. This technology is very much in the experimental stage and is currently being trialled at various Army bases across the US, but initial signs indicate the utility of the training. However, less than 43% of soldiers play videogames at least once a week (compared to 69% of civilians), which suggests there may be a need for "training for training," to enhance and speed up the benefits of virtual reality technology for military personnel (Orvis et al., 2010): if military personnel are not comfortable using this type of technology on a regular basis then the task demands associated with learning to use virtual reality equipment may offset the expected utility, and could result in training in incorrect habits and behaviours (Knerr, 2007).

Unfortunately, VEs can bring about severe motion sickness, resulting in nausea, headaches, dizziness, and severe disorientation. The primary cause is thought to be inconsistent body orientation, clashing with the motion received from the immersive headset (see cue conflict theory; Kolasinski, 1995). Between 20 and 40% of military pilots have been found to experience motion sickness following VE exposure (Kennedy et al., 1989), which is suprising because military pilots are typically less susceptible to motion sickness than the general population, as a direct result of their training. A Simulator Sickness Questionnaire, developed to measure adverse effects of VE training (SSQ; Kennedy et al., 1993) allows the quantification of adverse sickness effects on emerging technologies, but pre-exposure results also need to be compared for this to be an effective indication of adverse effects (Kolasinski, 1995). One concern with using this measure is that symptoms are not unique to motion sickness (they may also result from, for example, excessive alcohol, sleep deprivation, or the flu). However, the risks associated with motion sickness are thought to reduce with gradual exposure to the synthetic envionments over time (Knerr, 2007), and so controlled exposure in a safe (virtual) environment, irrespective of cause, is akin to exposure therapy, the benefits of which are likely to carry over to real world environments.

12.2.2 Treatment

Virtual imaging and immersive reality have begun to assist military patients with psychological illness and physical injuries that result from combat experiences. For example, burn-care patients revealed a 50% reduction in pain and anxiety ratings when immersed in a VE, compared to a video game distraction task (Hoffman et al., 2000a), with increased immersion being negatively correlated with pain reduction (Hoffman et al., 2000b). Immersion appears to decrease attention to painful stimuli, resulting in a reduced need for analgesia, and an overall improvement in the patient's tolerance to painful medical procedures (Shahrbanian et al., 2009).

Post-traumatic stress disorder (PTSD) is caused by traumatic events outside of typical human experience and is not uncommon in soldiers returning from combat and war zones. Graduated and prolonged exposure therapy, which is based on the assumption that repeated reliving and recounting of the traumatic event within a therapeutic setting will allow patients to manage their fears and memories, has well documented therapeutic outcomes for the treatment of PTSD (Rizzo et al., 2014). However, it relies on imagination and sensory memory, which can be problematic when individuals do not want to/cannot recount and verbalise experienced events (Rizzo et al., 2014). To overcome these problems, researchers have turned to VEs as an alternative means of therapy. Virtual Reality Exposure Therapy (VRET) provides a multisensory experience, enabling context-relevant cues to be produced using a VE to aid confrontation without effortful memory retrieval.

Therapists control the VRET to manipulate visual, olfactory, and audio cues presented, allowing for the presentation of specific and variable scenes relevant to each individual (Rizzo et al., 2008). McLay et al. (2012) found that of the 20 patients who completed VRET treatment, 75% no longer met the diagnostic criteria for PTSD, and that this improvement was still evident three months later for 76% of patients (also see Groves & Thompson, 1970; Thompson & Spencer, 1966). However, this research is in its early stages, typically has low sample sizes and high drop-out rates, and fails to pinpoint the locus of the reduction in PTSD symptoms.

The University of Southern California's Institute of Creative Technologies has started to create stress resilience training (STRIVE: STress Resilience in Virtual Environments; Rizzo et al., 2013), which is given to soldiers prior to deployment. This VE training aims to reduce PTSD symptoms and diagnoses for active military personnel by educating individuals to develop psychological coping skills for a variety of combat situations (for example, seeing fellow comrades injured from a roadside bomb, or watching innocent civilians dying). Throughout the training individuals get to know and interact with fellow military avatars, allowing immersive engagement to occur, akin to real-life relationships. Within each training episode, an emotionally challenging event occurs, developed from the feedback of service members diagnosed with PTSD. A virtual mentor then appears, guiding the individual through stress-reduction and management strategies, whilst additionally providing restructuring exercises to facilitate the process and appraisal of the event in a rational way, which is likened to a "digital emotional obstacle course." Allostatic load (AL) is the body's response to stress, which can be measured using physiological instruments such as pupil dilation and EEG (McEwen & Seeman, 1999). STRIVE is currently being used to investigate whether AL can predict individual resilience, and psychological and cognitive capability to experience traumatic events, often unavoidable within combat situations. Possibly in addition to pre-deployment training, this tool could be used as a psychological assessment pre-recruitment into the military?

There is stigma associated with seeking military psychiatric healthcare, possibly due to the assumption that mental illness may prevent personnel from progressing. SimCoach is a revolutionary approach to seeking advice on healthcare issues, designed especially for military personnel and their families by the US Defense Centers of Excellence for Psychological Health and Traumatic Brain Injury (DCoE). Consisting of customisable avatars, the virtual programme allows individuals to seek healthcare information anonymously, by interacting with embodied virtual interactive humans (avatars). Speech and even emotions are introduced via avatars whilst interacting with the "patient," providing confidential and anonymous feedback about the individual's history or clinical concerns (Rizzo et al., 2013; see https://www.youtube.com/watch?v=2bsMESwBeyg). Users interact with the avatar by typing text, to which they respond by speaking, providing subtitles, sending links and tips specific to each individual's needs, and creating surveys and questionnaires to allows the correct and unique advice to be given to each patient. Created by doctors, technicians, psychologists, and the experiences of military personnel, they emphasise that this system does not replace human experts or live care; instead it is a gateway to advice and support, and in certain cases, signposts users to relevant therapeutic real-life services.

12.3 Defence applications

12.3.1 Information-gathering and investigation

Gathering information is a fundamental goal for those concerned with protecting national and international security. One central challenge is understanding how to *move* people from withholding to imparting information. Additional challenges arise from the increasing use of synthetic/virtual reality environments as communication channels, and how to gather information when interacting in such environments. In recent years there has been a move away from coercive, interrogative interview methods towards intelligence interviewing (e.g. Intelligence Science Board, 2009; Janofsky, 2006; Wahlquist, 2010), which is information-gathering in context. Here, context includes physical, interpersonal, and informational environments, which intelligence interviewers should be cognisant of in order to develop a bespoke operational accord to maximise the possibility of information gain (see Boon et al., 2010; Intelligence Science Board, 2009).

Fundamental to developing an operational accord is "information power" – that is, *possessing* information about an interviewee (physical, interpersonal, and environmental) and *understanding how* to use that information to increase information gain. One example being the, a tactical approach to disclosing of information (Dando & Bull, 2011; Dando et al., 2015), or using information gained to influence the interviewees' perceptions and behaviour. Intelligence interviewing is grounded in social cognitive theory, which in brief argues that human cognition, that is the way in which humans "think," is a product of a reciprocal interplay between intrapersonal, behavioural, and environmental determinants

(Bandura, 1991). Accordingly, understanding the reciprocity between intrapersonal (internal to the communicator) aspects, and external environments offers exciting opportunities for understanding how to move to a position of information power.

To date, social cognitive theory and intelligence interviewing approaches have largely been FtF-centric, and so the question that arises is how might an operational accord be developed across different contexts? VEs offer exciting possibilities because they can be manipulated, and so can the way in which communicators represent themselves as avatars. Accordingly, they offer opportunities to discreetly collect information that can be used to "get to know" a receiver, to move to a position of "soft power" (Nye, 2004), whereby positive outcomes are possible without commanding them, and without having tangible power, but rather by affecting behaviour and shaping preferences. VEs may allow intelligence interviewers opportunities to do just that.

It is timely that consideration be given to VEs as interviewing spaces on several counts. First, there has been an exponential increase in our dependence on VEs: over 40% of the world's population currently have an Internet connection (compared to just 1% in 1995; www.internetlivestats.com), and cyberspace underpins national and international infrastructures (e.g., water, fuel, and banking). Access to VEs has resulted in increased crime and antisocial behaviours (identity theft, fraud, inciting hatred, sexual offending, harassment). Extremism and radicalisation has and is increasing in synthetic communication spaces (Cornish et al., 2009), with terrorist groups regularly using VEs to spread propaganda, raise funds, communicate, and plan attacks.

Currently, VEs are being developed and utilised for forensic and investigative training purposes, typically to simulate events and interactions to allow investigators to develop and practice skill sets more efficiently than might otherwise be that case, and to do so in a safe environment. For example, using an avatar-based interview simulator (ABIS) to allow free-flowing conversation, thus creating a realistic interactive training experience (Kuykendall, 2010). VEs could also be useful for harvesting information on a receiver's cognitive style, information that could then be used to support the development of an operational accord, and so affect information-gathering during an initial investigation and during more formal follow-up interviews (see Dando & Tranter, 2015).

Cognitive style is variously described, but in the main is a preferred method of managing specific cognitive tasks (Kozhevnikov et al., 2014; Zhang & Sternberg, 2006, 2009), and is believed to be *"stable attitudes, preferences, or habitual strategies that determine individuals' modes of perception, memory, thought and problem solving"* (Kozhevnikov et al., 2014, p. 4), which are environmentally sensitive (Buss & Greling, 1999). Hence, understanding cognitive styles may be the basis for developing a bespoke person-interaction fit. Cognitive styles have received much attention in the domains of education, business, and management, but as yet, despite the obvious application of models of cognitive style to investigative interviewing, there appears to be little empirical research in this domain.

VEs may countenance information power because human cognition and behaviour differs when communicating in SEs, compared to traditional FtF environments, and synthetic communication environments can be easily managed/manipulated to encourage the revelation/ collection of information. The psychological literature offers several hypotheses as to why behaviour in VEs differs compared to FtF. The online disinhibition hypothesis (Suler, 2004) suggests that individuals are increasingly willing to disclose more personal information online because the fantasy and invisibility elements of VEs allow communicators to remain anonymous. However, more recent research suggests that people do not always divulge more information. Rather, people are selective in the information that they share, typically offering extensive basic information in the first instance; only following some level of interaction do they divulge their innermost feelings and thoughts. Equally, other factors are believed to be as important as perceived disinhibition or perceived anonymity, such as knowing who one is talking to (e.g., the stranger on a train versus the known other effects), the content of self-revelations, and the mode of communication (people are more likely to divulge in 1:1 synchronous communications), for example (see Attrill, 2014).

Furthermore, there is a perception that the rules and regulations that govern in reality do not actually exist in VEs, because meaningful reprisal is extinguished from the conscious. The Equalisation hypothesis (Dubrovsky et al., 1991) supports this, arguing that VEs allow freedom from physical attributes such as race, gender, age, and physical disabilities, and so stereotypical behaviours that arise in traditional FtF interactions are not available in VEs. A key example of this comes from an early study by Matheson (1991), who used a negotiation task to manipulate the availability of gender cues. Social perceptions of gender were directly affected by the availability of this information. Gender stereotypical perceptions were absent until gender cues were revealed and became salient to participants, at which point women were perceived as more cooperative, and men as more exploitative, indicating that anonymity alters people's cognition, which in turn affects behaviour.

Positive affect is the instinctual reaction to positive, emotionally-provoking stimuli which can systematically influence performance on varying cognitive tasks without conscious awareness. One example is the International Affective Picture System (IAPS: Lang et al., 1999), which provides a set of normative, emotionally evocative pictures across a wide range of semantic categories. Implementing emotionally evocative backdrops within a synthetic environment offers possibilities for managing an environment to improve communication, and enhance cognition. Indeed, research does indicate that positive affect enhances problem-solving and decision-making as a result of more flexible, innovative, and efficient cognition (Isen, 2001). Positive affect has also been found to facilitate the bargaining process, improving outcomes when negotiating to buy and sell (Carnevale & Isen, 1986), apparently facilitating more systematic and careful processing of additional task information, and reducing distractibility and impulsivity. One avenue for future research is to consider integrating the IAPS into VEs to investigate cognition and positive effects for information-gathering.

The role of haptic feedback in collaborative tasks, that is whether haptic communication through forced feedback can facilitate a sense of being and collaboration with a remote partner, also speaks to intelligence gathering in VEs. Using multimodal shared virtual environments across gender and personality, simulating touch was found to have a powerful impact on task performance and sense of togetherness, which in turn affected cognitive processes, such as decision-making (e.g., Hafich et al., 2007). Making one subject "strong" and the other "weak" by way of a haptic device might offer environmental opportunities for information-gathering and criminal investigation.

One significant advantage of VEs for information-gathering is that they allow people to communicate as avatars. Avatars allow individuals to change aspects of their social identity to become less identifiable, or even create a novel, entirely fictitious and unrepresentative online identity – customising features such as eye colour, hair colour, height, gender, race, etc. It is believed that many people who use avatars online wish to be unique, different, and creative when immersed, allowing them to explore things they could not do in reality (Lin & Wang, 2014). However, others use avatars as an extension of their offline self, to explore aspects of their self that they are not overly confident with offline. Building confidence online allows them to then transfer those features of self offline. Others want to portray their actual self through avatars because they can obtain reinforcement and positive feedback for that self online that they might otherwise not receive offline – a form of acceptance (as one of the fundamental underlying needs of humans: social acceptance and belongingness).

Avatars have been found to influence cognition. For example, Yee and Bailenson (2007) found that when individuals were assigned an avatar, their cognition merged to this digital representation, changing their behaviour in accordance with the representation. This is referred to as the *Proteus Effect*, whereby people conform to the expectations and stereotypes of their given avatars altered self-representation, which has a direct effect on behaviour in VEs. For example, those assigned attractive avatars were found to display increased self-disclosure and were more willing to approach the opposite sex; and the taller the avatar, the more confident participants became when verbally communicating in the VE.

12.4 Linguistic behaviour

It has been suggested that a lack of media richness in VEs is a challenge for investigators that without being able to consider physical behaviour (often referred to as body language), alongside spoken and written verbal communication, senders may be less effective information gatherers, and are unlikely to make appropriate veracity judgments (Marett & George, 2004). However, recent research has suggested that trained investigators can be more effective in determining veracity FtF when considering only the informational content offered in reply to a sender's questions, rather than the paralinguistic and non-verbal cues commonly associated with deception (Dando & Bull, 2011; Dando et al., 2015; Jenkins & Dando, 2012). When communicating online, indicators of deceit are discernable in the complete absence of any physical behavioural cues simply by analysing language use (Tausczik & Pennebaker, 2010; Taylor et al., 2013). For example, use of words that denote distinctions and connections (e.g., but, also) can offer insights into the nature of people's reasoning, and interviewer initiated language matching between interviewer and interviewee has been associated with increased confessions (Richardson et al., 2014).

US law enforcement agencies are currently using a virtual reality simulation for training on interaction and communication styles within an interview setting (Kuykendall, 2010). The technology (known as SIMmersion) creates an avatar that allows trainees to practise their skills, thus reducing the need for costly classroom teaching. Police officers displayed much improved interviewing reaction time, response time, critical decision processes, and safety skills after completing training using this technology, findings that indicate the utility of SEs for allowing officers able to make mistakes, rewind, and practise their skills and techniques within a safe and secure environment.

12.5   Modelling threats

Researchers have just begun to investigate the utility of immersive gaming as a method for modelling and investigating threats. For example, modelling insider incidents (Dando et al., 2013) following the realisation that what is needed is a method for understanding insider behaviours, and the need to develop rapid investigative methods both to filter potential insider persons of interest, and to provide information relevant to planning an effective investigative strategy. Deceivers are known to attempt to control their verbal and physical behaviour when being interviewed about suspected wrongdoing, making veracity decisions difficult. Yet, computer-mediated VE communication by way of a triage interview has recently resulted in a high degree of success: veracity detection was more accurate (for truth-tellers and deceivers) because verbal and physical behaviour differences emerged online that were not apparent face-to-face (also see Dando & Bull, 2011; Dando et al., 2015).

Organisations concerned with serious and organised economic crime and counter-terrorism are increasingly turning to VEs to model occurrences associated with organised crime and terrorism. Projects include the development of pan-European VEs for data sharing, monitoring, and cooperative analysis, and understanding how best to use electronic movement data collected from simulation VEs to detect and predict terrorist occurrences in real life (e.g., Andrews et al., 2015; Sandham et al., 2015). Typically the literature reporting this type of modelling research is not openly available, for obvious reasons, and in any case this type of research is in its infancy (but see Stedmon & Lawson, 2015).

12.6   Conclusion and food for thought

Psychological research investigating the potential of VEs for military and defence innovation is timely in that it offers numerous interesting, and promising lines of enquiry, but as yet is not widely available. For example, group polarisation, which is the tendency for like-minded people to become extreme in their thinking following a group discussion, also occurs in virtual communities (McKenna & Green, 2002). Understanding the cognitive processes that support this phenomenon in VEs may prove beneficial for military and defence purposes. Knowing how judgements are formed and modified in VEs may allow the development of predictive models. Knowledge of mode of information processing style, namely intuitive-experiential or analytical-rational, might predict the likelihood of irrational behaviour in given certain circumstances (Denes-Raj & Epstein, 1994; Epstein & Pacini, 1999), which may offer methods for overriding rational cognitive systems to best effect. Understanding individual need for cognitive closure and how this affects behaviour in VEs would indicate whether individuals are more likely to "seize and freeze" upon initially presented information, and so close their minds to further knowledge, resulting in impulsive decision-making (Webster & Kruglanski, 1994). This style implies that individuals may be less likely to move from opposing to converging viewpoints, suggesting that intelligence gatherers need to be particularly cautious about how to manage initial approaches, at least in terms of the informational content of verbal interactions, perhaps?

Using virtual reality headsets to immerse participants in virtual worlds, manipulating environments, collecting information on immersed cognitive styles, and then measuring VE cognition compared to traditional face-to-face contexts, would further our understanding of ways to gather information in SEs (e.g., see Dando & Tranter, 2015; Tranter et al., 2014). The increasing number of individuals using online environments to communicate dictates that military and criminal investigators, and information gatherers *must* give serious consideration to the multiple contexts in which communication can occur – being proactive, rather than reactive, may reap significant rewards.

12.7 References

Andrews, S., Polovina, S., Akhgar, B., Stainforth, A., Fortune, D., & Stedmon, A. (2015). Tackling financial and economic crime through strategic intelligence management. In Stedmon, A. & Lawson (Eds.). *Hostile Intent and Counter-terrorism: Human Factors Theory and Application.* Ashgate Publishing, Ltd. Surrey, UK.

Attrill, A. (2014). The misconception of online splurging and associated security risks. *Cybertalk*, UK.

Bandura, A. (1991). Social cognitive theory of self-regulation. *Organizational Behavior and Human Decision Processes, 50,* 248–287.

Boon, K. E., Huq, A. Z., & Lovelace, D. C. (2010). *Assessing the GWOT* (Vol. 110). Oxford University Press, UK.

Buss, D., & Greling, H. (1999). Adaptive individual differences. *Journal of Personality, 67,* 209–243.

Carnevale, P. J. D., & Isen, A. M. (1986). The influence of positive affect and visual access on the discovery of integrative solutions in bilateral negotiation. *Organizational Behavior and Human Decision Processes, 37,* 1–13.

Cornish, P., Hughes, R., & Livingstone, D. (2009). *Cyberspace and the national security of the United Kingdom. Threats and responses.* London: Chatham House.

Cruz-Neira, C., Reiners, D., Springer, J. P., Neumann, C., Odom, C. N. S., & Kehring, K. (2011). *An integrated immersive simulator for the dismounted soldier.* Paper Presented at the Interservice/Industry Training, Simulation and Education Conference, Orlando, Floria.

Dando, C. J., & Bull, R. (2011). Maximising opportunities to detect verbal deception: Training police officers to interview tactically. *Journal of Investigative Psychology and Offender Profiling, 8,* 189–202.

Dando, C. J., Bull, R., Ormerod, T. C., & Sandham, A. L. (2015). Helping to sort the liars from the truth-tellers: The gradual revelation of information during investigative interviews. *Legal and Criminological Psychology, 20,* 114–128.

Dando, C. J., Taylor, P., & Ormerod, T. C. (2013). *Detecting deception: Can computers interview to detect persons of interest following an insider attack?* Paper Presented at the American Psychology and Law Conference, 7–9 March, Portland, Oregon, USA.

Dando, C. J., & Tranter, C. L. (2015). Intelligence interviewing: Synthetic environments, cognition and cognitive styles. *Investigative Interviewing: Research and Practise, 7,* 42–50.

Denes-Raj, V., & Epstein, S. (1994). Conflict between intuitive and rational processing: When people behave against their better judgment. *Journal of Personality and Social Psychology, 66,* 819–829.

Djaouti, D., Alvarez, J., Jessel, J. P., & Rampnoux, O. (2011). Origins of serious games. In *Serious games and edutainment applications* (pp. 25–43). London: Springer.

Dubrovsky, V. J., Kiesler, S., & Sethna, B. N. (1991). The equalization phenomenon: Status effects in computer-mediated and face-to-face decision-making groups. *Human-Computer Interaction, 6,* 119–146.

Eshel, T. (2011). Virtual reality prepares soldiers for the real fight. Retrieved from http://defense-update.com/20111031_virtual-reality-prepares-soldiers-for-the-real-fight.html#.VZqVbflViko

Fox, J., & Ahn, S. J. G. (2013). Avatars: Portraying, exploring, and changing online and offline identities. In *Handbook of research on technoself: Identity in a technological society* (pp. 255–271). Hershey, PA: IGI Global.

Gallagher, A. G., & Cates, C. U. (2004). Approval of virtual reality training for carotid stenting: What this means for procedural-based medicine. *Jama, 292,* 3024–3026.

Gehlbach, H., Young, L. V., & Roan, L. K. (2012). Teaching social perspective taking: How educators might learn from the Army. *Educational Psychology, 32,* 295–309.

Groves, P. M., & Thompson, R. F. (1970). Habituation: A dual-process theory. *Psychological review, 77,* 419–450.

Hafich, A., Fowlkes, J., & Lenihan, P. (2007, January). Use of haptic devices to provide contextual cues in a virtual environment for training. In *The Interservice/Industry Training, Simulation & Education Conference (I/ITSEC)* (Vol. 2007, No. 1). National Training Systems Association.

Hoffman, H. G., Doctor, J. N., Patterson, D. R., Carrougher, G. J., & Furness III, T. A. (2000a). Virtual reality as an adjunctive pain control during burn wound care in adolescent patients. *Pain, 85,* 305–309.

Hoffman, H. G., Patterson, D. R., & Carrougher, G. J. (2000b). Use of virtual reality for adjunctive treatment of adult burn pain during physical therapy: A controlled study. *The Clinical Journal of Pain, 16,* 244–250.

Intelligence Science Board. (2009). *Intelligence interviewing*. National Intelligence Agency, USA.

Isen, A. M. (2001). An influence of positive affect on decision making on complex situations: Theoretical issues with practical implications. *Journal of Consumer Psychology, 11,* 75–85.

Janofsky, J. S. (2006). Lies and coercion: Why Psychiatrists should not participate in police and intelligence interrogations. *The Journal of the American Academy of Psychiatry and the Law, 34,* 472–478.

Jenkins, M. C., & Dando, C. J. (2012). Computer-mediated investigative interviews: A potential screening tool for the detection of insider threat. In S. Tomblin, N. MacLeod, R. Sousa-Silva, & M. Coulthard (Eds.). *Proceedings of the 10th Biennial Conference of the International Conference of Forensic Linguistics, Birmingham: Centre for Forensic Linguistics* (pp. 272–282).

Kennedy, R. S., Lane, N. E., Berbaum, K. S., & Lilienthal, M. G. (1993). Simulator sickness questionnaire: An enhanced method for quantifying simulator sickness. *The International Journal of Aviation Psychology, 3,* 203–220.

Kennedy, R. S., Lilienthal, M. G., Berbaum, K. S., Baltzley, D. R., & McCauley, M. E. (1989). Simulator sickness in US Navy flight simulators. *Aviation, Space, and Environmental Medicine, 60,* 10–16.

Knerr, B. W. (2007). *Immersive simulation training for the dismounted soldier* (Report No. ARI-SR-2007-01). Army research Institution Filed Unit, Orlando, Florida.

Kolasinski, E. M. (1995). *Simulator sickness in virtual environments* (Report No. ARI-TR-1027). Army Research Institute for the Behavioural and Social Sciences, Alexandria, VA.

Kozhevnikov, M., Evans, C., & Kosslyn, S. M. (2014). Cognitive style as environmentally sensitive individual differences in cognition: A modern synthesis and applications in education, business, and management. *Psychological Science in the Public Interest, 15,* 3–33.

Kuykendall, J. (2010). Future training innovations. *FLETC Journal, 8,* 9–12.

Lang, P. J., Bradley, M. M., & Cuthbert, B. N. (1999). *International affective picture system (IAPS): Instruction manual and affective ratings.* The Center for Research in Psychophysiology, University of Florida.

Lathan, C. E., Tracey, M. R., Sebrechts, M. M., Clawson, D. M., & Higgins, G. A. (2002). Using virtual environments as training simulators: Measuring transfer. In *Handbook of virtual environments: Design, implementation, and applications,* Taylor &Francis: CRC Press (pp. 403–414).

Lee, A. (2006). Flight simulation, virtual environments in aviation. *Aviation, Space, and Environmental Medicine, 77,* 164.

Lin, H., & Wang, H. (2014). Avatar creation in virtual worlds: Behaviors and motivations. *Computers in Human Behavior, 34,* 213–218.

Macedonia, M. (2002). Games soldiers play. *IEEE Spectrum, 39,* 32–37.

Manojlovich, J., Prasithsangaree, P., Hughes, S., Chen, J., & Lewi, M. (2003). Utsaf: A multi-agent-based framework for supporting military-based distributed interactive simulations in 3d virtual environments. *Proceedings of the 2003 Winter Simulation Conference,* New Orleans (Vol. 1, pp. 960–968).

Marett, K., & George, J. F. (2004). Deception in the case of one sender and multiple receivers. *Group Decision and Negotiation, 13,* 29–44.

Matheson, K. (1991). Social cues in computer-mediated negotiations: Gender makes a difference. *Computers in Human Behavior, 7,* 137–145.

McEwen, B. S., & Seeman, T. (1999). Protective and damaging effects of mediators of stress: Elaborating and testing the concepts of allostasis and allostatic load. *Annals of the New York Academy of Sciences, 896*, 30–47.

McKenna, K. Y., & Green, A. S. (2002). Virtual group dynamics. *Group Dynamics: Theory, Research, and Practice, 6*, 116.

McLay, R. N., Graap, K., Spira, J., Perlman, K., Johnston, S., Rothbaum, B. O., Difede, J., Deal, W., Oliver, D., Baird, A., Bordnick, P. S., Spitalnick, J., Pyne, J. M., & Rizzo, A. (2012). Development and testing of virtual reality exposure therapy for post-traumatic stress disorder in active duty service members who served in Iraq and Afghanistan. *Military medicine, 177*, 635–642.

Nye, J. S. (2004). *Soft power: The means to success in world politics.* PublicAffairs.

Orvis, K. A., Moore, J. C., Belanich, J., Murphy, J. S., & Horn, D. B. (2010). Are soldiers gamers? Videogame usage among soldiers and implications for the effective use of serious videogames for military training. *Military Psychology, 22*, 143–157.

Pacini, R., & Epstein, S. (1999). The relation of rational and experiential information processing styles to personality, basic beliefs, and the ratio-bias phenomenon. *Journal of Personality and Social Psychology, 76*, 972–989.

Richardson, B. H., Taylor, P. J., Snook, B., Conchie, S. M., & Bennell, C. (2014). Language style matching and police interrogation outcomes. *Law and Human Behavior, 38*, 357.

Rizzo, A., Buckwalter, J. G., Forbell, E., Reist, C., Difede, J., Rothbaum, B. O., Lange, B., Koenig, S., & Talbot, T. (2013). Virtual reality applications to address the wounds of war. *Psychiatric Annals, 43*, 123–138.

Rizzo, A., Hartholt, A., Grimani, M., Leeds, A., & Liewer, M. (2014). Virtual reality exposure therapy for combat-related posttraumatic stress disorder. *Computer, 47*, 31–37.

Rizzo, A., Pair, J., Graap, K., Rothbaum, B. O., Difede, J., Reger, G., & Perlman, K. (2008). Virtual Iraq: Initial results from a VR exposure therapy application for combat-related post traumatic stress disorder. *Medicine Meets Virtual Reality, 16*, 420–425.

Rose, F. D., Attree, E. A., Brooks, B. M., Parslow, D. M., & Penn, P. R. (2000). Training in virtual environments: Transfer to real world tasks and equivalence to real task training. *Ergonomics, 43*, 494–511.

Sandham, A., Ormerod, T., Dando, C., & Menacere T. (2015). On the trail of the terrorist: A research environment to simulate criminal investigations. In Stedmon, A. & Lawson (Eds.). *Hostile Intent and Counter-terrorism: Human Factors Theory and Application.* Ashgate Publishing, Ltd. Surrey, UK.

Shahrbanian, S., Ma, X., Korner-Bitensky, N., & Simmonds, M. J. (2009). Scientific evidence for the effectiveness of virtual reality for pain reduction in adults with acute or chronic pain. *Annual Review of Cyberpsychology and Telemedicine, 144*, 40–43.

Suler, J. (2004). The online disinhibition effect. *Cyberpsychology & Behavior, 7*, 321–326.

Tausczik, Y. R., & Pennebaker, J. W. (2010). The psychological meaning of words: LIWC and computerized text analysis methods. *Journal of Language and Social Psychology, 29*, 24–54.

Taylor, P. J., Dando, C. J., Ormerod, T. C., Ball, L. J., Jenkins, M, C., Sandham, A., & Menacere, T. (2013). Detecting insider threats through language change. *Law and Human Behavior, 37*, 267–275.

Teteris, E., Fraser, K., Wright, B., & McLaughlin, K. (2012). Does training learners on simulators benefit real patients? *Advances in Health Sciences Education, 17,* 137–144.

Thompson, R. F., & Spencer, W. A. (1966). Habituation: A model phenomenon for the study of neuronal substrates of behavior. *Psychological Review, 73,* 16–43.

Tranter, C., Dando, C. J., & Sandham, A. (2014).*Cognition in context: Investigating the impact of individual differences on information-gathering interviews in synthetic environments.* Paper Presented at the International Investigative Interviewing Research Group Annual Conference, 2–4 June, Lausanne, Switzerland.

Wang, Z. C., Tsai, C. C., & Chien, M. C. (2012). Design of an intelligent soldier combat training system. *International Journal of Automation and Smart Technology, 2,* 309–317.

Wahlquist, J. A. (2010). Setting and agenda for the high-value detainee interrogation group. *International Journal of Intelligence Ethics, 1,* 24–45.

Webster, D. M., & Kruglanski, A. W. (1994). Individual differences in need for cognitive closure. *Journal of Personality and Social Psychology, 67,* 1049–1062.

Wilson, C. (2008). *Avatars, virtual reality technology, and the U.S. military: Emerging policy issues* (Report No. RS22857). Washington, D.C.

Witmer, B. G., & Singer, M. J. (1998). Measuring presence in virtual environments: A presence questionnaire. *Presence: Teleoperators and Virtual Environments, 7,* 225–240.

Yee, N., & Bailenson, J. N. (2007). The Proteus Effect: Self transformations in virtual reality. *Human Communication Research, 33,* 271–290.

Zhang, L. F., & Sternberg, R. J. (2006). *The nature of intellectual styles.* Psychology Press, Mahwah, New Jersey: London.

Zhang, L. F., & Sternberg, R. J. (2009). Revisiting the value issue in intellectual styles. *Perspectives on the Nature of Intellectual Styles,* 63–85.

Ziv, A., Wolpe, P. R., Small, S. D., & Glick, S. (2003). Simulation-based medical education: An ethical imperative. *Academic Medicine, 78,* 783–788.

13
Social Media Impact on Organisations

Johanna Myddleton and Chris Fullwood

13.1 Introduction

There is little doubt that information technology has permanently altered the manner in which we work. We are now living in a digital age, where working practices are no longer constrained by space and time. Co-workers are not necessarily required to be in the same physical environment to work together and this has led to an evolution in business practices, with organisations increasingly managing larger workforces spread over different time-zones (Bennett et al., 2010b; McGregor, 2000). One of the many challenges faced by organisations who adopt distributed ways of working comes with building and maintaining an organisational culture and ensuring that employees feel part of the team and work cooperatively, despite the geographical distance from their co-workers. Distributed ways of working may sometimes engender feelings of isolation, reduce productivity, and lower staff morale (Bell et al., 2008; Bennett et al., 2010b). However, although technology may in this case be contributing to the problem, it could also provide the solution. Increasingly, organisations are introducing social media tools into the workplace which have the potential to improve communications between co-workers, boost morale and motivation, increase efficiency, and foster an organisational culture (Akkirman & Harris, 2005; Bennett et al., 2010b). Moreover, we might also anticipate similar advantages for employees who work co-presently. Despite these prospective benefits however, there are various ramifications for the marriage of social media and the world of work which need to be addressed (e.g. inappropriate use), as these will likely impact policy around social media use in the workplace as well as the manner in which organisations make effective use of these tools.

This chapter aims to provide a broad discussion of the different ways in which organisations may utilise social media to improve communications between employees and as a tool to foster more effective interaction with customers and stakeholders. The potential benefits of social media for organisations and workers will be outlined, whilst many of the pitfalls that may be involved in its use will also be highlighted. The authors acknowledge that this is an expansive subject and will, where possible, offer suggested reading for topics that have not been covered in depth. Particular focus will be given to the impact of social media on workplace productivity, the different ways in which companies can use these tools to build brand-awareness and reputation management, and to the manner in which social networking sites can help employees to form closer bonds with colleagues and to share knowledge within an organisation. The authors will also offer recommendations based on a review of the extant theory and literature for more effective use of what could arguably be one of the most significant technological advances for the business world this century.

13.2   Social media and social capital

Social media are Web 2.0 technologies which facilitate interactive communications between individuals and groups, publically and privately, and where the content is controlled by the contributors rather than the site owner (Kaplan & Haenlein, 2010). Social media tools include blog sites, photo-sharing applications, social gaming sites and social networking sites (SNSs), which offer the added facility to view and browse the social connections of other members (Edosomwan et al., 2011). Although there are a wide variety of social media tools, this chapter will focus almost exclusively on the impact of SNSs on organisations, due to their rapid uptake by many companies (Skeels & Grudin, 2009). The majority of SNSs share core features and tend towards the maintenance of pre-existing relationships and strengthening of latent social ties (Ellison et al., 2007). What is unique to SNSs is the ability to manifest a discernible list of existing social ties within the site and to communicate with these individuals publically, privately, individually, or collectively (boyd & Ellison, 2007). Whilst for casual users this may simply translate into a convenient tool for sharing news with family and friends simultaneously, SNSs offer employees a space for networking beyond the boardroom, scouting for career progression opportunities, and enhancing personal relationships with colleagues.

Whilst many companies actively encourage the use of pre-existing SNSs (e.g. Facebook) (Fraser & Dutta, 2008), others create SNSs for sole use by their employees. Research suggests the use of SNSs can increase employee morale and create a more positive perception of the organisational culture. One particular study created an SNS called "Beehive" for IBM employees, which was hosted on the company's intranet server. It offered the same functionalities as "public" SNSs, including the ability to connect with other employees, create a profile, and share user-created content such as photos and lists, but lacked the private, synchronous messaging that SNSs such as Facebook allow. Interestingly, members suggested that they used the internal SNS to actively strengthen latent ties (potential relationships which have not been fully developed yet) with new and loosely-bonded colleagues (DiMicco et al., 2008). In this way, by reciprocally self-disclosing increasingly personal information and communicating more frequently online with colleagues, they experienced increased feelings of closeness and strengthened their relationships; subsequently increasing their *social capital*. Social capital has been defined as the resources gained via any type of relationship existing between people; this may be physical or monetary but importantly also includes psychological and emotional resources (Coleman, 1988). Those with high levels of social capital will feel more fulfilled, wanted, and valued by their community (Baumeister & Leary, 1995). It is therefore not surprising that increased social capital relates to improved psychological well-being, higher self-esteem, and general satisfaction with life (Bargh & McKenna, 2004; Helliwell & Putnam, 2004). Given that more satisfied and valued members of a workforce are more productive and have higher organisational commitment (Harrison et al., 2006; Levy, 2003), encouraging workers to strengthen bonds within the company should accrue benefits on both a personal and organisational level. This was demonstrated in the results of a similar study by Koch et al. (2012), who found that an internal SNS resulted in positive emotions in new staff and reduced entry level staff turnover, as they felt welcome, supported, and part of a workplace family. Newly hired members of staff received higher performance ratings at review and were seen to be more efficient and effective, following quicker acclimatisation than previous cohorts who had not used the system (Koch et al., 2012).

As well as enabling employees to increase contact with loose ties within an organisation (which would arguably take more resources to manage offline), SNSs can also strengthen pre-existing bonds (Steinfield et al., 2009). In this way, SNSs may potentially benefit colleagues and companies in a twofold manner: 1) by increasing the number of connections and therefore ideas, opinions, and resources each colleague has access to; which allows them to seek out expertise and work collaboratively to overcome problems (Fraser & Dutta, 2008), whilst at the same time decreasing demand on immediate supervisors, who would normally be required to assist in this process (Koch et al., 2012); 2) by increasing the level of emotional support between colleagues, individuals will also feel better able to complete their own tasks, and the negative health impacts of workplace stress such as depression, stress-related absence, and anxiety may be lessened (Koch et al., 2012; LaRocco et al., 1980).

Whilst the positive impacts of social capital gained through SNSs have not yet been studied in great depth, research findings indicate that organisational citizenship behaviours, organisational commitment, and increased knowledge-sharing contribute to a positive organisational culture, which in turn ultimately leads to better performance of the company as a whole, whether it is a profit-driven or non-profit organisation (Arnold et al., 2010). In particular, knowledge-sharing and corporate citizenship are more prevalent in workplaces where supportive relationships between colleagues are encouraged and social capital levels are increased (Arnold et al., 2010). As such, if SNSs can be used to foster these relationships, they offer the potential to have a largely positive impact on companies.

13.3   Social media for knowledge exchange

The flow of knowledge in large and international companies can be severely disrupted due to barriers such as geography, time-zone differences, departmentalisation, and hierarchy. Additionally, when companies group individuals by skill sets, it can be difficult to access specialist knowledge or alternative solutions to problems (Williamson, 2009). SNSs may provide a way for companies to increase knowledge flow between all areas of the company by allowing employees an increased support system of variable knowledge and skill sets (Bennett et al., 2010a). In addition, utilising social media such as blogs, Twitter, and wikis can promote the sharing of knowledge from individuals to many employees simultaneously, as well as an online space for asking and answering questions and creating collaborative documentation (Williamson, 2009). A number of companies, such as Microsoft, successfully use blogs during product development to share ideas and invite feedback, and this shared creation is also thought to help build a strong project team identity. Similarly, Nike runs a public blog which details product development for customers (Lee & Trimi, 2008). The use of these tools may help form an organisational culture of knowledge-sharing, as individuals will be able to share in a way that may expend less energy and resources. For example, by creating an online FAQ, organisations can avoid the need to respond to multiple queries via e-mail or phone. Workers will also be able to see the impact of their contribution through elements such as "liking," comments, and other feedback, and this will help create an online prestige and reputation which rewards employees with personal recognition and will therefore motivate them to contribute further (van Zyl, 2009). Innovative ideas from ground level employees may also skip a few steps in the company ladder, which may be harder to achieve offline (Bennett et al., 2010a). This has obvious advantages, in that it allows for a more streamlined and quicker approach to problem-solving, however it may result in resentment for some middle-management (DiMicco et al., 2008).

In one study, the use of an internal SNS left middle-management feeling jealous and somewhat aggrieved in response to younger and newer staff having online interactions with CEOs and gaining knowledge and acknowledgement that they may never have been able to achieve within the confines of a "traditional" offline work environment. As such, these staff felt disadvantaged and that entry level employees needed to "do their time" in order progress within the company, in the same way that they had to (Koch et al., 2012). This feeling of injustice is concerning, as it could lead to a loss of organisational commitment and lower job satisfaction, therefore negatively impacting on job performance (LaRocco et al., 1980). Despite this, however, we must also consider that in the coming years it will be the "Net generation" who predominately fill these middle-management positions. In this event, these initial feelings of dissent may be phased out, as social networks and a more linear interaction between staff members becomes the norm. Managers who have seen the benefits of the inbuilt rewards of such a collaborative and non-hierarchical online system will move into positions of power, and attitudes are therefore likely to shift.

Whilst the advantages of increased knowledge exchange and collaborative working environments appear obvious (e.g. faster problem-solving, increased innovation, less time creating duplicate solutions where staff may be unaware of parallel projects – all of which may increase profits), what is less clear is how to involve all members of the organisation in this process through the use of SNSs, rather than just those entering the company at ground level. A positive aspect of SNSs is that they may encourage contributions from those who lack confidence (Ardichvilli et al., 2003), due to the greater feeling of community and sense of disinhibition frequently felt online (see Suler, 2004). Indeed, online interactions are thought to be more accessible to those with higher levels of anxiety, due to reduced fear of rejection and increased time to control the construction of posts (Joinson, 2004). For instance, those who feel too anxious to contribute during a face-to-face meeting may feel more comfortable posting online, as they will not be able to see the expressions of others, who they fear may respond negatively to their ideas. However, companies should offer clear guidelines on what should and should not be posted online, as the online disinhibition effect also tells us that individuals might be more aggressive or anti-social, or share inappropriate information in online spaces due to a reduction in social presence and social cues (Suler, 2004).

Using SNSs for knowledge-sharing also offers the ability for short asynchronous (not requiring the communicator and receiver to both be online at the same time) contributions, which can reach a wide audience and be accessed at virtually any time or from any location. In addition, real-time person-to-person and group discussions can take place, which may be advantageous to those whose roles are highly demanding in terms of time and resources. This also means more employees of varying degrees of seniority and skill level may be able to contribute and share mentoring of others when time to do so is becoming increasingly short (DiMicco et al., 2008). Whilst this form of knowledge-sharing offers clear benefits to individuals and organisations alike (Bennett et al., 2010a), companies should also consider the security risks involved. Employees should be well versed on what is and is not appropriate to send via open networks, how to protect confidential information and trade secrets, and the ways in which posts online may impact on the reputations and privacy of individuals and organisations (Wang & Kobsa, 2009). One way to overcome these issues is to create an internal SNS behind the firewalls of the organisation's intranet (DiMicco et al., 2008). However, by taking this route, companies are restricting the network to those within the company and therefore limiting the skill set and knowledge base available.

13.4   Social media use and workplace productivity

Whilst promoting the use of SNSs in the workplace may potentially create benefits in terms of increasing social capital, productivity, and knowledge-sharing, there exists a strong possibility that some employees may misuse these sites and abuse their privileges. Many organisations restrict access to or completely prohibit the use of social media during working hours, which may stem from fears that engaging with SNSs will distract employees from workplace tasks. Evidence suggests that between 34% (Young & Case, 2004) and 60% (Greenfield & Davis, 2002) of organisations report either disciplining or terminating the contracts of employees for Internet use for non-work reasons. Further, one study by Nucleus Research (2009) found that nearly two thirds of employees spent on average 15 minutes of the work day on Facebook, which is thought to equate to a 1.5% loss in productivity across the working year. This raises two issues for employers: 1) that social media may in fact lead to a decrease in productivity, and 2) that attempts to prevent this may be futile. Let's look at each of these in turn.

Whilst the Nielson (2009) report indicated a "loss" in productivity of 1.5% over the year, what it failed to distinguish was whether this time spent on Facebook was in addition to other forms of work avoidance or instead of, meaning that it is unclear whether it could be considered an actual loss to the company. Secondly, participants were not asked what they were doing whilst on Facebook, meaning that conceivably some of this time online could have been spent conversing with colleagues or seeking information. Indeed, it seems that the younger generation in particular are turning to Facebook as an alternative medium for interacting and conversing with colleagues during work hours, for work purposes (DiMicco et al., 2008). However, alternative evidence also suggests that younger individuals may be more likely to log on to Facebook as a means to procrastinate as a means of avoiding work-related tasks (Orchard et al., 2014). Either way, it is unclear whether time spent on these social media really equates to loss of productivity, or whether it is simply another tool for work. The answer to this may depend on the user themselves as well as the organisational culture in the company they work for.

In addition to the company's concerns, SNS use at work may also create an additional aspect of role conflict for employees who try to separate their personal and working lives. As SNS members will be friends with family and friends as well as colleagues, those who are unable to balance these two roles may experience mental strain, stress, and negative emotions. On the other hand, some employees may find it easier to maintain familial relationships whilst at work, which may help to reduce role conflict (Bennett et al., 2010a). Additionally, we should acknowledge that the individual has the right not to join a social media site if this is their wish. A study by Orchard et al. (2015) suggests that some individuals might feel compelled to use sites like Facebook through peer pressure, despite experiencing membership in an almost entirely negative light. For example, some people feel that Facebook membership actually isolates them from their offline friends and that it breeds paranoia due to increased opportunities for surveillance. Forcing these individuals to join SNS sites for company purposes may therefore result in resentment and unnecessary stress. At the same time, if an organisation encourages SNS use but does not make it mandatory, non-members may feel left out of the loop or even disadvantaged in terms of the opportunities that may become available for working their way up the organisational ladder. Organisations may therefore wish to consider alternative, more "traditional" methods for employees to gain recognition and interact with senior management to ensure fairness and equal opportunities for all. SNSs should not replace more traditional modes of communication, but should be used to supplement them.

These individual differences in attitudes to social media were also highlighted in a study by O'Driscoll and Cummings (2009). It was noted that older programmers found access to social media at work a distraction, whereas younger employees in the same department suggested that social media were an important creative tool in their arsenal and that removing access to these sites would diminish productivity. Banning social media altogether may have an impact on productivity by lowering staff morale. Prohibiting access to the Internet, or oppressive surveillance of social media may leave employees feeling as if they cannot be trusted and lead to lower levels of perceived autonomy and independence, which are essential components of human well-being (see Baard et al., 2004 for a discussion of self-determination theory and the positive effects of autonomy in the workplace). This in turn has been seen to decrease job satisfaction, lower feelings of self-efficacy, and increase intention to leave, all of which have been associated with lower productivity, particularly given that self-efficacy impacts on the amount of effort people put into tasks and their perseverance when faced with difficulties (Levy, 2003). Whilst to some employers, preventing access to social media may seem like a trivial issue, or even necessary to prevent workers wasting valuable time, for employees entering the working arena for the first time, who are used to accessing SNSs at any time desired through mobile technology, this may feel overly restrictive and a contributory factor to creating a domineering organisational culture. In particular, where open office spaces and hot-desking restrict employees' ability to personalise their own distinct work space, being able to access a SNS where individuals have control over their profile and the ability to express themselves may be an important factor in developing and protecting their workplace identity and sense of autonomy (Bennett et al., 2010a).

Another area where allowing access to social media may be able to prevent loss of productivity is in roles which require the repetition of simple tasks. For example, O'Driscoll and Cummings (2009) found that by allowing entry level IT staff to interact and plan social activities with colleagues during work hours, they professed the work environment was "fun," despite claiming their actual roles were monotonous and boring. This light relief meant that whilst staff were perhaps not working for all the hours they were compensated for, the quality of their work may have been higher and staff retention rates for this department increased, reducing hiring and training costs for the company, which may balance out any costs of time spent socialising during work hours. In this way, allowing social media for personal use at work may need to be considered as somewhat of a cost-benefit trade-off. Indeed, Coker (2011) found that short, frequent instances of workplace Internet leisure browsing (WILB), which did not exceed 12% of the working day, actually enhanced productivity. Short, frequent breaks have been found to restore concentration and enhance motivation and creativity. People have also been shown to *over*compensate for time spent on non-work related tasks by working harder and faster when returning to work, meaning overall output is higher (Zijlstra et al., 1999). That said, for tasks that require high levels of cognitive resources or extended concentration, notifications popping up from Facebook or extended social media "breaks" may indeed interrupt work-flow and negatively impact on productivity. With little ability to control time on social media if these sites are not blocked in the workplace, we are again returned to the idea of companies restricting use completely in order to prevent the potential overuse of social media by employees. Banning social media completely may also limit security risks, such as downloading malware and viruses, and stop social media using valuable bandwidth data and slowing the network (for a discussion on how organisational policy may be able to mediate the effects of social media use at work, see Wickramasinghe & Nisaf, 2013).

In spite of organisational goals, however, banning SNSs or Internet use completely for staff may be somewhat of a futile and counterproductive practice. A middle ground may be to allow access for finite periods of time (e.g. "Facebook Fridays"; Lee et al., 2007) for personal use (also, see Moqbel et al., 2013 for a working model and more detailed discussion of how personal social network use can increase job performance). This allows the company some element of control over use and articulates a clear policy for staff members on what is acceptable use of company time and SNSs – an important factor cited throughout the literature in regards to protecting company reputation and preventing a legal backlash should evidence from SNSs be used against staff in disciplinary procedures. So far the discussion has focused on the relationship between employees and their use of SNSs in the workplace. We will now shift our attention to the manner in which organisations can use SNSs to raise brand-awareness and engage with their customers.

13.5   Raising brand-awareness with social media

When typing your company name into Google, what do you see? Perhaps the company website comes first, most likely followed by Wikipedia, news stories, Facebook, and Twitter accounts. All excepting the first are, to a certain degree, beyond the company's immediate sphere of influence and demonstrate the powerful standing that user-generated content and social media now have. To a large extent, PR, marketing, and company reputation has switched from a unidirectional monologue directed *at* consumers *by* professionals, to a consumer-contrived, interactive conversation that will occur with or without the contribution of companies (Jones et al., 2009). What may particularly concern companies is when this conversation is negative and damaging to the organisation's reputation and therefore profits. For example, following a BBC *Panorama* programme that highlighted the use of child labour in Primark's supply chain, this became a top news story which dominated the top hits of Google searches, and the Twittersphere was alive with pejorative comments that were tweeted and retweeted across the globe (Jones et al., 2009). Disenchanted customers and employees too are turning to social media to share their views and experiences, but more importantly, others are listening.

One easy way to see this in action is to look at product reviews in comparison with sales figures. Quite simply, the better the reviews, the more sales are made (Chevalier & Mayzlin, 2006). A survey by Nielson (2012) found that online reviews and advice are the second most trusted source of information on products (after recommendations by family/friends), which in turn has an impact on purchasing decisions. Recently however, research has suggested that not only are people influenced in their purchasing, but their opinions of products and the services they have used are also socially influenced. People rate previously positively-reviewed products higher than they would without seeing others' feedback, creating a positive bias towards products (Arai, 2013). Given that on sites such as Amazon, reviewers are not required to purchase a product prior to reviewing, this potentially gives companies an opportunity to manipulate sales by posting positive reviews and allowing this positive bias and the bandwagon effect (see Chapter 10) to boost sales and inflate feedback.

It also seems to be the case that negative feedback has a far more substantial impact on business, compared to positive feedback. People frequently rely on mental shortcuts, such as the availability heuristic, to reduce cognitive load during decision-making. One drawback of this is that people are often swayed by personal stories of negative experiences that are more salient to them at the time they make their decision (Gilovich et al., 2002). For instance, when purchasing a car, if you know someone who had a particularly bad experience with a Volkswagen, you might avoid this manufacturer despite its reliability according to independent statistical data. Equally, if you are researching a company online and the top hits of a Google search are negative news stories, you may rely on this immediately available data rather than searching for further information. This makes it particularly vital that when organisational crises occur they are dealt with swiftly and appropriately, and that using social media like Twitter to adopt an apologetic tone can be considerably more effective than press releases and information posted on blogs. "Rescuing" a negative reputation is beyond the scope of this chapter, however see Schultz et al. (2011) for guidance on how this may be managed, and see Warin et al. (2013) for an account of how game theory can be utilised for enhancing and protecting reputation using social media. Regardless of a company's presence (or absence) on social media, it is clear that consumers will continue these conversations. For many "prosumers" (consumers who actively research and are highly knowledgeable about products and services), company SNS profiles are now one of the first ports of call for gathering information (Clarke, 2008), and organisations with a social media presence are seen as more technologically advanced and "in touch" with customers than those without (Ventola, 2014). Some individuals even report that they would choose which hospital to attend based on social media presence (Huang & Dunbar, 2013). So rather than simply sitting back and letting the conversation happen without them, can companies utilise social media for purposive marketing?

13.6　Social media as an advertising tool

Globally, companies of all sizes are recognising the potential for incorporating social media into their marketing strategies, and logically it make sense. The number of people joining and regularly using social media is steadily increasing across all age groups and demographics, and as people are spending more time online, they are spending less time watching television and listening to the radio (Gillin, 2007). From this, it is clear that the audience for classic "ad break" campaigns may be decreasing, whilst the potential online audience is increasing substantially (Gillin, 2007). Add to this that marketing online via social media can be seen as requiring fewer resources (in terms of time, manpower, and financial cost), and it is clear why so many companies are reducing their marketing budgets elsewhere and jumping online instead (Kirtis & Karahan, 2011). It seems currently that organisational adoption of social media is far outstripping research on the implications of its use (Raeth et al., 2009). Regardless of this lack of information, Michaelidoue et al. (2011) found that over a quarter of small and medium business-to-business companies surveyed had included some form of social media in their current marketing strategy, with 44% planning to increase their budget in this area, and it is thought that the proportion of business-to-consumer companies using SNSs is substantially higher. However, less than half the companies surveyed assessed the success of SNS marketing campaigns and anecdotal evidence from consumers themselves seems to suggest that not only does it not always draw in new customers or improve customer relations, but that advertising on SNSs may in fact have a negative impact on branding (Diffley et al., 2011; Michaelidoue et al., 2011).

Diffley et al. (2011) found that SNS users found the majority of adverts on Facebook and other SNSs to be irrelevant, irritating, and an invasion of personal space. Advertisements in SNSs are often guided by previous browsing; for example, shopping for a new pair of shoes on one site may result in an advert for the same shoes when moving on to another. Pop-ups, flashing images, and roll-over adverts were mistrusted by users, who feared they might contain viruses and resented that they interrupted their current task. It was very rare that users clicked on adverts on purpose, and when they did, they often became frustrated that it took them away from the main site. When adverts were followed it tended only to be by well-known brands that were trusted by users and when they already had an established awareness of the company (Diffley et al., 2011). As people are confronted with adverts in almost all aspects of the online world, people frequently use heuristics (mental shortcuts or "rules of thumb") to reduce the cognitive strain that would be required to decide whether each website or message is legitimate or safe to use (Metzger et al., 2010). As such, brands which are well established and popular may be viewed as more trustworthy, as people might assume that "if everyone is using it, it must be ok" or "X must be good because so many people like it." The same can be said for brands known from other forms of advertising, such as television adverts, which are viewed as being more stringently monitored by authorities and therefore less likely to be making false claims. Similarly, companies and consumers may rely on "trustmarkers" such as the "padlock" symbol in the address bar when accessing a website, which gives a visual representation about the safety and privacy of the connection to the site. Symbols like this work in much the same way the "kite" symbol signifies quality on merchandise and tend to result in a transference of trust. In other words, people assume that if an authoritative source which they trust certifies a company, then that company must be just as trustworthy (Aiken & Boush, 2006). However, it may also be relatively easy for sites to mimic these symbols, and whilst individuals can check the authenticity of the trust symbols, for many the presence of one, in conjunction with a professional looking website, may be enough to garner trust (Aiken & Boush, 2006).

The tendency to follow only brands known from other areas of life (e.g. television marketing, previous use of products), which seems to stem from people having more trust in offline advertising and known brands, is completely at odds with organisational goal of gaining new customers through social media adverts. In addition, it contradicts perceptions that presence on social media alone will create a more dynamic and interactive relationship with customers. In actual fact it seems consumers only add companies with whom they already have an affiliation to their SNSs, and are quite adept at and insistent on ignoring advertising within SNSs (Diffley et al., 2011). However, this is not to say that they couldn't work; it is simply that – as we are finding across many areas of cyberpsychology – what works offline cannot automatically be transferred online with the same outcome. What does seem to work online, however, is "word of mouth."

13.7   Word of mouth

Word of mouth has long been recognised as a useful tool in the marketing kit for raising brand-awareness and attracting new customers, and social media now allows that "word," good or bad, to reach a wider network of people. When surveyed, the majority of people stated that online recommendations were valuable, credible, and honest, and around 49% of individuals said that information found on social media sites influenced their purchase decision (DEI Worldwide, 2008). Early research also seems to suggest that recommendations made by online connections are likely to attract more and longer-term customers (Trusov et al., 2008). But how do companies stimulate people to talk about and recommend their company to others? As we have seen, direct advertising does not seem to be well received, and so more indirect measures may be required. Companies such as Coca-Cola have utilised consumer-made viral videos for this purpose. Viral videos are those that spread quickly through networks via the sharing facility and tend towards product placement in interesting, shocking, or humorous videos, rather than direct advertising. In this way the brand is associated with the "buzz" created by the video, without forcing itself on consumers, who reject company intrusion into their online social life (Diffley et al., 2011).

An alternative solution is to incentivise individuals by offering discounts, coupons, free samples, and other rewards when they share online. Whilst this has shown some success, companies should be aware that *requiring* customers to share with others in order to gain access to information or basic privileges may have the opposite effect (Diffley et al., 2011), which again relates to removing control over their personal use of social media. So it would seem then that whilst organisations may be able to utilise social media for marketing purposes, the techniques used need to be placed more in line with the way that consumers use these sites. Individuals see their profiles, blogs, and accounts as their own personal space, where they create the content and over which they wish to remain in control. As such, companies need to be "invited" into this personal space and recognise the switch from "consumers" to "prosumers" (Davis & Moy, 2007). One way to do this is via word of mouth, by encouraging and incentivising current customers to share their positive experiences with their network of friends and associates. Again this does not necessarily mean that these new customers will interact in the way hoped, so perhaps, rather than mass targeting, it may be more beneficial to reach the smaller proportion of individuals who are active users, willing to converse with companies and who can raise profiles on the company's behalf. Indeed, some companies have already picked up on this idea. For instance, cosmetics firms may send beauty bloggers with large followings the latest products free of charge, in the hope of a good review and increased sales (Chaovalit, 2014). What is clear however is that significantly more research needs to be done in this area before companies invest in the "wrong" type of social media marketing and that above all, companies should aim for transparency and honesty in their online presence.

Whilst this section has focused on improving performance for profit-driven organisations, it is also worth noting that charity and non-profit organisations were some of the first to utilise social media for organisational and marketing purposes. SNSs and social media have been used to interact with volunteers and donors, educate about programmes and services, and develop relationships with stakeholders and the public (Waters, 2009). SNSs are particularly useful for consumer relationship building, given their interactive nature, and research in this area encourages non-profits to engage users by: providing external links to news and websites which demonstrate current work, advertising opportunities for offline involvement, such as volunteering and events, and asking users to contribute actively, e.g. giving their e-mail addresses and donating online (Waters, 2009). Currently however, many organisations are underutilising SNSs or creating organisational profiles without a designated strategy for improving relations with the public/consumers, which somewhat limits the organisational benefits of these sites (Waters, 2009).

13.8 Summary and conclusions

One area which we have not covered in this chapter relates to the ways in which those seeking employment may make use of social media to enhance their job prospects and, conversely, how poor online impression management may adversely impact upon employability. This is currently an under-researched area, but worth briefly introducing at this stage to highlight that further research is needed here. For example, evidence suggests that organisations are increasingly screening job candidates pre-interview, via social media, to gain additional information on them which may not be communicated in a CV or covering letter (Davison et al., 2012). Because the online world encourages people to communicate in a more grammatically loose and informal fashion (e.g. with the inclusion of more textspeak, colloquialisms, and slang), this may impact on how favourably some individuals are perceived. For instance, in one experiment people rated those who used textspeak as being less intelligent and less employable than those who used grammatically correct language (Scott et al., 2014). Moreover, textspeak users are perceived as less conscientious and scoring lower in openness, and therefore may be deemed more "sloppy," "unprofessional," and "unimaginative" (Fullwood et al., 2015), which may harm employment prospects. As previously discussed, the online disinhibition effect (Suler, 2004) may also result in users over-sharing, and one may only speculate as to the impact this might have on potential employers' perceptions (e.g. if SNS users bad-mouth the company they work for or share too much personal and private information, which may demonstrate a lack of restraint and poor judgement). It is not just organisations, therefore, who should be aware of reputation management. Individuals should also be very wary about indiscriminately using textspeak in their profiles and over-sharing some types of information, and consider the implications that this might have for them in the job market (Fullwood et al., 2015).

The impact of social media in the working environment is an expansive and ever-growing area of academic interest. As we have noted in this chapter, social media can be used as a powerful and cost-effective tool to improve working relations, strengthen bonds, and share knowledge within an organisation, particularly for companies who use distributed ways of working. At the same time, we have highlighted how access to social media during working hours has the potential to decrease productivity. Although prohibiting access to social media is likely to be counterproductive (e.g. because it may reduce morale and restrict the stress-reducing, knowledge-sharing, and social capital benefits of employee involvement), clear policies for appropriate use, written in easy to understand language and enforced consistently, should be communicated to employees. We have also seen how social media can be utilised as an effective tool to raise brand-awareness and for reputation building, while at the same time noting the pitfalls associated with direct advertising strategies via social media. Utilising word of mouth promotion through incentivising consumers with rewards for sharing and recommending pages and products may be more fruitful than overt advertising tactics. Social media are not going away and, whether we like it or not, are becoming ever more closely connected to our working lives. Even if organisations are not yet ready to embrace social media fully, it would be imprudent to completely overlook the important role that they could potentially play in improving communication between employees, and with partners, stakeholders, and customers.

13.9 References

Aiken, K. D., & Boush, D. M. (2006). Trustmarks, objective-source ratings, and implied investments in advertising: Investigating online trust and the context-specific nature of Internet signals. *Journal of the Academy of Marketing Science, 34*, 308–323.

Akkirman, A. D., & Harris, D. L. (2005). Organizational communication satisfaction in the virtual workplace. *Journal of Management Development, 24*(5), 397–409.

Arai, S. (2013). The problem with online ratings. *MIT Sloan Management Review, 55*(2), 47–52.

Ardichvilli, A., Page, V., & Wentling, T. (2003). Motivation and barriers to participation in virtual knowledge-sharing communities of practice. *Journal of Knowledge Management, 7*(1), 64–77.

Arnold, J., Randall, R., Patterson, F., Silverster, J., Robertson, I., Cooper, C., Burnes, B., Harris, D., Axtell, C., & Hartog, D. D. (2010). *Work psychology: Understanding human behaviour in the workplace* (5th Ed.). Italy: Prentice Hall.

Baard, P. P., Deci, E. L., & Ryan, R. M. (2004). Intrinsic need satisfaction: A motivational basis of performance and wellbeing in two work settings. *Journal of Applied Social Psychology, 34*, 2045–2068.

Bargh, J. A., & McKenna, K. Y. A. (2004). The Internet and social life. *Annual Review Psychology, 55*, 573–590.

Baumeister, R. F., & Leary, M. R. (1995). The need to belong: Desire for interpersonal attachments as a fundamental human motivation. *Psychological Bulletin, 117*(3), 497–529.

Bell, A., Graham, R., Hardy, B., Harrison, A., Stansall, P., & White, A. (2008). *Working Without Walls*. London: OGC and DEGW.

Bennett, J., Owers, M., Pitt, M., & Tucker, M. (2010a). Workplace impact of social networking. *Property Management, 28*(3), 138–148.

Bennett, J., Pitt, M., & Owers, M. (2010b). Workplace impact of social networking. In W. Atherton & B. Saghafi (Eds.), *Proceedings of the 5th Annual Bean Conference (Liverpool Conference on the Built Environment and Natural Environment)* (pp. 60–72). Liverpool: Liverpool John Moores University.

boyd, d. m., & Ellison, N. B. (2007). Social network sites: Definition, history, and scholarship. *Journal of Computer-Mediated Communication, 13*(1), 210–230.

Chaovalit, P. (2014). Factors influencing cosmetics purchase intention in Thailand: A study on the relationship of credibility and reputation with in persuasive capabilities of beauty bloggers. *AU-GSB e-Journal, 7*(1), 34–42.

Chevalier, J., & Mayslin, D. (2006). The effect of word of mouth on sales: Online book reviews. *Journal of Marketing Research, 43*(3), 345–354.

Clarke, R. (2008). Web 2.0 as syndication. *Journal of Theoretical and Applied Electronic Commerce Research, 3*, 30–43.

Coker, B. L. S. (2011). Freedom to surf: The positive effects of workplace Internet leisure browsing. *New Technology, Work and Employment, 26*(3), 238–247.

Coleman, J. S. (1988). Social capital in the creation of human capital. *The American Journal of Sociology, 94*, 95–120.

Davis, C., & Moy, C. (2007). *Coming to Terms with Business Transparency*. Admap Magazine, 487, Oxon: WARC, 19–22.

Davison, H. K., Maraist, C. C., Hamilton, R. H., & Bing, M. N. (2012). To screen or not to screen? Using the Internet for selection decisions. *Employee Responsibilities and Rights Journal, 24*, 1–21.

DEI Worldwide. (2008). *Engaging Consumers Online: The Impact of Social Media on Purchasing Behavior*. Retrieved on 11 August 2015 from https://themarketingguy.files.wordpress.com/2008/12/dei-study-engaging-consumers-online-summary.pdf

Diffley, S., Kearns, J., Bennett, W., & Kawalek, P. (2011). Consumer behaviour in social networking sites: Implications for marketers. *Irish Journal of Management, 30*(2), 47–65.

DiMicco, J., Millen, D. R., Geyer, W., Dugan, C., Brownholtz, B., & Muller, M. (2008). Motivations for social networking at work. In *Proceedings of the 2008 ACM conference on computer supported cooperative work* (pp. 711–720). ACM: New York.

Edosomwan, S., Prakasan, S. K., Kouame, D., Watson, J., & Seymour, T. (2011). The history of social media and its impact on business. *The Journal of Applied Management and Entrepreneurship, 16*(3), 79–91.

Ellison, N. B., Steinfield, C., & Lampe, C. (2007). The benefits of Facebook 'friends': Social capital and college students' use of online social network sites. *Journal of Computer-Mediated Communication, 12*, 1143–1168.

Fraser, M., & Dutta, S. (2008). *Throwing Sheep in the Boardroom: How Online Social Networking Will Transform Your Life, Work and World.* John Wiley: West Sussex.

Fullwood, C., Quinn, S., Chen-Wilson, J., Chadwick, D., & Reynolds, K. (2015). Put on a smiley face: Textspeak and personality perceptions. *Cyberpsychology, Behavior and Social Networking, 18*(3), 147–151.

Gilovich, T. D., Griffin, D., & Kahneman, D. (2002). *Heuristics and Biases: The Psychology of Intuitive Judgment.* New York, NY: Cambridge University Press.

Gillin, P. (2007). *The New Influencers: A Marketer's Guide to the New Social Media.* Sanger, CA: Quill Driver Books/Word Dancer Press, Inc.

Greenfield, D. N., & Davis, R. A. (2002). Lost in cyberspace: The web @ work. *CyberPsychology, Behavior and Social Networking, 5*, 347–353.

Harrison, D. A., Newman, D. A., & Roth, P. L. (2006). How important are job attitudes? Meta-analytic comparisons of integrative behavioural outcomes and time sequences. *Academy of Management Journal, 49*(2), 305–325.

Helliwell, J. F., & Putnam, R. D. (2004). The social context of well-being. *Philosophical Transactions of the Royal Society B: Biological Sciences, 359*(1449), 1435–1446.

Huang, E., & Dunbar, C. L. (2013). Connecting to patients via social media: A hype or a reality? *Journal of Medical Marketing, 13*(1), 14–23.

Joinson, A. N. (2004). Self-esteem, interpersonal risk, and preference for e-mail to face-to-face communication. *CyberPsychology, Behavior and Social Networking, 7*(4), 472–478.

Jones, B., Temperley, J., & Lima, A. (2009). Corporate reputation in the era of Web 2.0: The case of Primark. *Journal of Marketing Management, 25*(9), 927–939.

Kaplan, A. M., & Haenlein, M. (2010). Users of the world, unite! The challenges and opportunities of social media. *Business Horizons, 53*, 59–69.

Kirtis, K., & Karahan, F. (2011). To be or not to be in social media arena as the most cost efficient marketing strategy after the global recession. *Procedia Social and Behavioural Sciences, 24*, 260–268.

Koch, H., Gonzalez, E., & Leidner, D. (2012). Bridging the work/social divide: The emotional response to organizational social networking sites. *European Journal of Information Systems, 21*(6), 699–717.

LaRocco, J. M., House, J. S., & French, J. R. P. (1980). Social support, occupational stress and health. *Journal of Health and Social Behaviour, 21*(3), 202–218.

Lee, Y., Lee, Z., & Kim, Y. (2007). Understanding personal web usage in organizations. *Journal of Organizational Computing and Electronic Commerce, 17*, 75–99.

Lee, S., & Trimi, S. (2008). Organizational blogs: Overview and research agenda. *International Journal of Information Technology and Management, 7*(2), 113–119.

Levy, P. L. (2003). *Industrial/Organizational Psychology: Understanding the Workplace.* Boston: Houghton Mifflin Company.

McGregor, W. (2000). The future of workspace management. *Facilities, 18*(3/4), 138–143.

Metzger, M. J., Flanagin, A. J., & Medder, R. B. (2010). Social and heuristic approaches to credibility evaluation online. *Journal of Communication, 60*, 413–439.

Michaelidoue, N., Siamagka, N. T., & Christodoulides, G. (2011). Usage, barriers and measurement of social media marketing: An investigation of small and medium B2B brands. *Industrial Marketing Management, 40*, 1153–1159.

Moqbel, M., Nevo, S., & Kock, N. (2013). Organizational members' use of social networking sites and job performance: An exploratory study. *Information Technology & People, 26*(3), 240–264.

Nielson. (2009). *Global Faces and Networked Places: A Nielsen Report on Social Networkings New Global Footprint.* Retrieved on 11 August 2015 from http://www.nielsen.com/us/en/insights/reports/2009/Social-Networking-New-Global-Footprint.html

Nielson. (2012). *State of the Media: The Social Media Report.* Retrieved on 11 August 2015 from http://www.nielsen.com/us/en/insights/reports/2012/state-of-the-media-the-social-media-report-2012.html

Nucleus Research. (2009). *Facebook: Measuring the Cost to Business of Social Networking.* Report J57. Retrieved on 11 August 2015 from http://nucleusresearch.com/research/single/facebook-measuring-the-cost-to-business-of-social-notworking/

O'Driscoll, T., & Cummings, J. (2009). Moving from attention to detail to detail to attention: Web 2.0 management challenges in Red Hat and the Fedora project. In Proceedings of the SIM Enterprise and Industry Application for Web 2.0 Workshop. In G. Kane, A. Majchrzak, and B. Ives (Eds.), *Society for Information Management.* Phoenix, AZ.

Orchard, L., Fullwood, C., Morris, N., & Galbraith, N. (2015). Investigating the Facebook experience through Q methodology: Collective investment and a 'Borg' mentality. *New Media and Society.* 17(9), 1547–1565.

Orchard, L., Fullwood, C., Galbraith, N., & Morris, N. (2014). Individual differences as predictors of social networking. *Journal of Computer Mediated Communication, 19*(3), 388–402.

Raeth, P., Smolnik, S., Urbach, N., & Zimmer, C. (2009). Towards assessing the success of social software in corporate environments. *AMCIS 2009 Proceedings.* Retrieved on 11 August 2015 from http://aisel.aisnet.org/amcis2009/662

Schultz, F., Utz, S., & Gorits, A. (2011). Is the medium the message? Perceptions of and reactions to crisis communication via twitter, blogs and traditional media. *Public Relations Review, 37,* 20–27.

Scott, G. G., Sinclair, J., Short, E., & Bruce, G. (2014). It's not what you say, it's how you say it: Language use on Facebook impacts employability but not attractiveness. *Cyberpsychology, Behavior and Social Networking, 17*(8), 562–566.

Skeels, M. M., & Grudin, J. (2009). When social networks cross boundaries: A case study of workplace use of Facebook and LinkedIn. In S. Teasley and E. Havn (Eds.), *Proceedings of the GROUP'09* (pp. 95–104). ACM: New York.

Steinfield, C., DiMicco, J. M., Ellison, N. B., & Lampe, C. (2009). Bowling online: Social networking and social capital within the organization. In *Proceedings of the Fourth International Conference on Communities and Technologies* (pp. 245–254). ACM: New York.

Suler, J. (2004). The online disinhibition effect. *Cyberpsychology, Behavior and Social Networking, 7,* 321–326.

Trusov, M., Bucklin, R. E., & Pauwels, K. (2008). Effects of word-of-mouth versus traditional marketing: Findings from an Internet social networking site. *Journal of Marketing, 73*(5), 90–102.

Van Zyl, A. S. (2009). The impact of Social Networking 2.0 on organisations. *The Electronic Library, 27*(6), 906–918.

Ventola, C. L. (2014). Social media and health care professionals: Benefits, risks, and best practice. *Pharmacy and Therapeutics, 39*(7), 491–499.

Wang, Y., & Kobsa, A. (2009). Privacy in online social networking at workplace. *Computational Science and Engineering, 4,* 975–978.

Warin, T., Marcellis-Warrin, N., Sanger, W., Nembot, B., & Mirza, V. H. (2013). Corporate reputation and social media: A game theory approach. *International Journal of Economics and Business Research, 9*(1), 1–22.

Waters, R. D. (2009). Engaging stakeholders through social networking: How non-profit organizations are using Facebook. *Public Relations Review, 35,* 102–106.

Wickramasinghe, V., & Nisaf, M. S. M. (2013). Organizational policy as a moderator between online social networking and job performance. *VINE, 43*(2), 161–184.

Williamson, B. (2009). Managing at a distance. *Business Week,* 27 July.

Young, K. S., & Case, C. J. (2004). Internet abuse in the workplace: New trends in risk management. *Cyberpsychology, Behavior and Social Networking, 7,* 105–111.

Zijlstra, F. R., Roe, R. A., Leonora, A. B., & Krediet, I. (1999). Temporal factors in mental work: Effects of interrupted activities. *Journal of Occupational and Organizational Psychology, 72*(2), 163–185.

14
Online Psychometric Assessment

Daniel Hinton and Debbie Stevens-Gill

14.1 Psychometric assessment

At its core, the field of psychometrics is concerned with the measurement of psychological constructs. The term *psychometric* is derived from the ancient Greek words ψυχικός ("of the soul"; "of life") and μέτρησις ("measurement"), and describes a group of methods by which a psychologist can measure a test taker's cognitive ability, personality, attitudes, interests, or other psychological characteristics relevant to a wide variety of therapeutic, occupational, educational, and forensic settings. These measurements are based on the test taker's responses to a series of questions and statements, known as *items*, traditionally administered using a pencil-and-paper system of question booklets and answer sheets. Within practitioner circles (as is the case in this chapter), "psychometrics," "psychological assessment," and "psychological measurement" are terms that are used interchangeably (Coaley, 2014).

The pioneering work of early psychologists such as Sir Francis Galton, Charles Spearman, and Hans Eysenck explored the structure of these constructs, leading to the development of modern theories of individual differences such as Carroll's (1993) Hierarchical Model of Intelligence, and the Big Five Model (Goldberg, 1990), which is respected today as the dominant trait model of the structure of personality. The design of all modern psychometric measures is informed by these foundational theories, seeking to tap into the hidden constructs they describe through interpretation of a test taker's responses to their items.

A key characteristic of psychometric measures is *standardisation*. The instructions provided to test takers, the environment in which they complete a measure, and the way in which their responses are scored are all standardised. This ensures that – as far as is practicable – all test takers should have the same experience of completing the same psychometric, irrespective of when, where, or for what purpose it is completed, and that the random error associated with measuring these hidden constructs is minimised.

The use of psychometrics has permeated every discipline of applied psychology to some extent. The main reason that psychometric measures are used so extensively by practitioners is that they are associated with a number of important outcomes. Cognitive ability tests are one of the best predictors of future job performance available in job selection (Robertson & Smith, 2001; Schmidt & Hunter, 1998). Additionally, they can accurately predict one's likelihood of success on training programmes (Ackerman et al., 1995; Ree & Carretta, 1998; Ree & Earles, 1991), academic success in higher education (DeBerard et al., 2004; Harackiewicz et al., 2002), and have been linked to seemingly unrelated constructs such as one's political ideology (Hodson & Busseri, 2012). Personality inventories can be used to predict how attracted to an organisation a job applicant is likely to be (Lievens et al., 2001), how motivated and satisfied an employee will be in their work (Furnham et al., 2009), and even the type of career a child will pursue in his or her adult life (Woods & Hampson, 2010). The degree to which a person's work interests match the environment in which they work – as measured by Career Interest Inventories – is linked to important intrinsic factors such as their job satisfaction, but also extrinsic ones such as salary and degree of progression within a career (Bretz & Judge, 1994). Finally, clinical diagnostics allow practitioners in forensic settings to make assessments of risk, such as prediction of violent behaviour and self-harm among offenders (Morey & Quigley, 2002). Considered together, psychometric measures represent a powerful set of tools for practitioners in a vast range of specialisms.

As the prevalence of psychometrics continues to grow, so too does their potential for misuse. Therefore, in the UK, use of psychometrics is strictly governed by the British Psychological Society's Psychological Testing Centre. In order to purchase and use any psychometric measure, a test user must hold a qualification relevant to the specific context in which they intend to use testing. The qualifications currently offered by the BPS in testing are Test User: Occupational (Ability; Personality), Test User: Educational, and Test User: Forensic Contexts.

14.2 Online psychometrics

With the development and proliferation of Information Technology, traditionally paper-based organisational processes have gradually begun to be replaced by computer-based equivalents (Stone et al., 2015). Psychometric testing is no exception, the physical question booklets and answer sheets of traditional psychometric measures gradually having been replaced by computer-based assessment platforms. In recent years, test publishers have brought an increasingly diverse range of assessment tools to market, some of which are designed to replace existing paper-based tools directly, while others improve upon them, offering features that were previously impossible (or, at least, very costly).

Online psychometrics offer a number of advantages over traditional ones. Firstly, the process of scoring test takers' responses can be automated. This confers two advantages, in that measurement precision can be improved by the removal of human error in calculating standardised scores, and in that the scoring process can be substantially sped up, potentially allowing a practitioner to score thousands of responses in seconds, something that might take many hours to do had the tests been administered in a paper-based format. Related to this, the generation of feedback reports is instant for online measures. This dramatically decreases the time required to generate these reports, as they can – particularly in the case of personality feedback – be labour-intensive to produce by hand. Finally, web-based psychometrics can be remotely administered to test takers. This has had the knock-on effect of providing test publishers with much larger data samples – from a much wider range of cultural and geographical backgrounds – than has previously been practicable. For example, conversion of the *Implicit Association Test* – a very well-known measure that is most frequently used to assess implicit attitudes towards different groups of people – to an online format has allowed Harvard-based social psychology group *Project Implicit* to collect data from more than a million test takers (Gosling & Mason, 2015). Remote administration, therefore, has given publishers the potential to develop their measures iteratively to assess candidates with greater accuracy, to be more representative of specific demographic groups, and to assess whether their measures function in the same way cross-culturally and across national borders.

In addition to these general advantages, the way in which each broad class of measure is used has been transformed by the transition to online formats. The remainder of this section will describe some of these innovations and the impact that they have had on practice.

14.3   Cognitive ability tests

Many traditional ability tests have made the transition from paper-based formats to online. The *Raven's Progressive Matrices* series (Pearson

Education, 2015a), the *Wonderlic Personnel Test* (Wonderlic, 2014), and the newest iterations of the *Wechsler Adult Intelligence Scales* (WAIS; Pearson Education, 2015b) and *Wechsler Intelligence Scales for Children* (WISC; Pearson Education, 2015c) are all now offered in online formats designed to replicate, as far as possible, their offline counterparts.

In addition to these direct translations, the move to computer-based ability testing has brought with it the potential for novel approaches to the assessment of ability. A prime example of one such approach that would be impossible without the introduction of computer-based assessment platforms is *Computerised Adaptive Testing* (CAT). Rather than presenting candidates with a static, unchanging set of test items, CATs draw on a large pool of items to tailor the test's content to each candidate in real-time, based on their previous responses. In their most common form, candidates who answer a particular question correctly are presented with a harder question, whereas those who answer the same question incorrectly are shown an easier one.

The test terminates once enough response data has been collected for its algorithm to accurately estimate the candidate's true level of ability. The advantage of this approach is that it optimises the set of test items that make up the CAT for each test taker (van der Linden, 2008), so CATs tend to give a more reliable assessment of ability – in fewer items – than their static counterparts. Beyond this, CATs have been developed that can control for an individual applicant's speed when responding to test items (van der Linden, 2009), further increasing the accuracy of ability estimations. Prototype CATs have been developed for mobile devices (Triantafillou et al., 2008), giving the potential for their more flexible deployment in a variety of contexts in the future. Finally, the dynamic nature of CATs means that they are less susceptible to practice effects, and provide less opportunity for candidates to gain an unfair advantage through familiarisation with the test prior to completion (Kantrowitz et al., 2011).

14.4 Personality inventories

Similarly to ability testing, personality assessment is becoming increasingly popular online, particularly as part of recruitment and selection, training, career guidance, and personnel development. There is some evidence to suggest that online versus face-to-face measures of personality measures yield similar results (Bjornsdottir et al., 2014; Chuah et al., 2006). Personality inventories broadly fall into one of two categories: those that measure personality *traits* and those that measure personality *type*.

Trait measures, such as the *NEO PI-R* (Costa & McCrae, 1992), and the *Sixteen Personality Factors Questionnaire* (16PF; Conn & Rieke, 1994), measure personality as a set of relatively stable and enduring characteristics that describe the individual's preferred patterns of behaviour. These traits represent continua of opposing behavioural styles. Traits have been shown to predict performance at work (Penney et al., 2011). For example, trait conscientiousness can predict meeting sales success when combined with trait extroversion and self-efficacy (Yang et al., 2011). Personality traits can also be predictive of health outcomes. They have been shown to predict problematic drinking behaviours (Mezquita et al., 2010), as well as how susceptible employees are to burnout (Armon et al., 2012).

By contrast, type measures, such as the extremely popular Myers-Briggs Type Indicator (MBTI; Myers et al., 1998), classify respondents into one of a number of discrete personality types. Though this makes them attractive in that they are easily understood by test takers, they are not as flexible as trait measures, and tend to provide less detailed descriptions of personality. For these reasons, they are unsuitable for use in high-stakes situations, such as job selection. However, they are useful tools for team development and career guidance contexts, and there is good empirical evidence that personality type can predict psychological outcomes. For example, those classified as having a personality type characterised by a tendency towards negative emotions tend to display greater susceptibility to job stressors and job strain (Spector & O'Connell, 1994).

The increasing popularity of online personality measures is partly due to their ease of implementation and that there are fewer drawbacks to online personality assessment compared to online ability testing. Foremost among the advantages of using online personality assessment is that the testing environment does not need to be as strictly controlled as in ability testing. Furthermore, Buchanan (1999) proposed that – due to the perceived anonymity the Internet provides – online personality assessment might encourage a more honest portrayal of one's self than traditional methods do. To support this, Gosling et al. (2011) have highlighted the accuracy of people's portrayal of their personalities online, suggesting that they use social networking sites to extend their offline personalities (see also Back et al., 2010).

Adaptive testing has also found some application in online personality assessment. Adaptive personality measures have the advantage of dramatically reducing the time needed to assess a candidate's personality, an attractive benefit, given that these tools typically take much longer to administer than ability tests do. However, Ortner (2008) observed that changing the order in which personality measures are presented (as is the case in adaptive personality measures) can lead to differences in how personality is assessed across candidates.

In spite of the strengths that online ability and personality measures can bring, many organisations find them somewhat generic, preferring instead to approach assessment using bespoke measures that can be tailored to specific contexts. One type of psychometric measure that offers the potential for this kind of customisation is the Situational Judgement Test.

14.5   Situational Judgement Tests

Though it has existed as a concept for almost 100 years (Bergman et al., 2006), one form of psychometric measure that has only recently become popular is the Situational Judgement Test (SJT). SJTs present the test taker with a series of hypothetical situations and a number of pre-determined responses to that situation, instructing them to select the response that best represents either what they *would* do (for behaviour-based SJTs) or what they *should* do (for knowledge-based SJTs) were they in that situation themselves. Based on their responses, SJTs can be used to make inferences about traits within a test taker as diverse as their integrity, empathy, and interpersonal skills (Lievens et al., 2005).

Classically, these situations were presented to the candidate as text. However, increased bandwidth has allowed online SJTs to replace the text-based situations of their predecessors with those presented to candidates through video streaming. These more high-fidelity SJTs tend to be rated as more job-relevant by candidates when used for selection (Chan & Schmitt, 1997), and can predict future job performance more accurately than text-based ones (Christian et al., 2010). Furthermore, they eliminate performance differences attributable to differences in reading comprehension, and tend to display smaller ethnic group performance differences than traditional SJTs do (Chan & Schmitt, 1997). Saville Consulting, a leading UK-based psychometric test publisher, provides an interesting demonstration of the versatility that online SJTs can offer in one of their promotional videos (https://youtu.be/gIrhmC_NDHc).

14.6   Issues with online psychometrics

14.6.1   Pseudoscientific measures

Arguably the single greatest threat to the use of online psychometrics is the proliferation of web-based services posing as psychometric measures, but which are based on pseudoscience. Social media are awash with links to quiz sites, some contributors to which have realised – quite reasonably – that the public have an interest in self-discovery. Seeking to capitalise on this, these contributors have designed quizzes that promise to uncover the secrets of the taker's personality or to measure their IQ as accurately as a professional general ability test, all presented in a quick, friendly, colourful and – above all – fun format. Though these quizzes are often entertaining, the danger of them is that there is no barrier to entry for creation of this kind of content as there is for psychometrics, so they are unlikely to have been designed with the same scientific rigour as true psychometrics are.

In the case of quizzes that purport to measure personality type or traits, many rely on the use of Barnum Statements when describing a test taker's personality. Barnum Statements are statements which the majority of people will view as an accurate description of themselves. Though named after businessman and renowned hoaxer P.T. Barnum, the first empirical work by a psychologist on this phenomenon was conducted by Bertram R. Forer (1949). In Forer's landmark study, he administered a short personality questionnaire to a group of his psychology students. Having completed the questionnaire, each student was provided with some feedback, which Forer told them was tailored to their responses. In actual fact, though, Forer had given each student the *exact same feedback*, word for word. The students were then asked to rate the accuracy of their personality profile. Astonishingly, the students' average rating of the accuracy of their particular feedback was found to be over 85%. This effect has been demonstrated again and again, notably by entertainer Derren Brown. For a segment of his television series, Brown produced a "personality summary" similar to Forer's. So convincing was this summary that, on reading it, a member of his crew became convinced that it had been written for him, and that he was the target of a practical joke (Brown, 2006).

Barnum Statements tend to be presented in one of two forms. The first are statements of universal validity, broad enough as to be applicable to anyone, for example *"Security is one of your major goals in life"* (Forer, 1949, p. 120). By contrast, the second claim that one will feel or behave in one way some of the time, but, in other situations, they will react in the opposite way, for example *"At times you are extroverted, affable, sociable, while at other times you are introverted, wary, reserved"* (Forer, 1949, p. 120). As Forer observed, these statements are seductive: they feel like they describe our own individual personalities in intimate detail. However, they provide no differentiation between test takers, thus do not provide feedback of any value.

Therefore, online "pop" personality quizzes should only be used for the purposes of entertainment. It is often not apparent – particularly to those unfamiliar with psychometrics – how online quizzes differ from true psychometric measures. For this reason, these pseudoscientific measures represent a threat to the credibility of psychometrics.

14.7 Distortion and faking

A recurrent problem for some psychometric tests – particularly personality measures and behaviour-based SJTs – is the opportunity that candidates have to distort the impression that they create of themselves. This phenomenon – called, variously, Impression Management (IM), Socially Desirable Responding behaviour (SDR), or, simply, faking – is an unavoidable facet of self-report measures such as these. While it is very difficult to fake intelligence, it is far less challenging to make out, when asked, that one is more conscientious than is actually the case. This potentially becomes a serious problem in high-stakes situations such as job selection, where the temptation is strong for a candidate to appear more desirable to an employer than they are in reality.

This situation could, potentially, be exacerbated by the remote administration of online measures, given that they are unsupervised. This issue is one that has troubled practitioners for some time, particularly as a growing body of evidence emerges of how individuals engage in IM in other online activities, such as on social networks. Rosenberg and Egbert (2011) suggest that individuals engage in IM tactics online, with tactics becoming more prominent for certain personality traits. In the context of recruitment and selection, the role of IM tactics in online psychometric testing for employment purposes may be similar to that already observed in job interviews (Weiss & Feldman, 2006).

In spite of these concerns, empirical evidence is emerging that the incidence of SDR is not significantly higher for personality measures administered remotely than in supervised conditions (Arthur et al., 2010). This goes some way to building confidence in their use. However, the possibility of faking is an issue that continues to be problematic for practitioners using these tools, Peeters and Lievens (2005) observing that, when candidates fake their responses, a tool's potential to predict future job performance falls to around a third of what it is when they respond truthfully.

Over time, test publishers have developed a number of approaches to addressing this issue. One common method is to insert an SDR scale into the items of the questionnaire. A scale such as this is made up of items to which a positive response appears desirable, but which, in practice, almost no-one would endorse. For example, in a job selection situation one might be tempted to endorse the item *"I have never been late for an appointment"* to create a good impression for a potential employer, even though the likelihood of this being the case is extremely slim. A high score on an SDR scale would indicate that – across the whole tool – the candidate has deliberately tried to appear more desirable than they are, so would throw into question the validity of their personality profile. Alternatively, Sjöberg (2015) suggests that if SDR scales are used to statistically correct personality scale scores, around 90% of the effects of faking behaviour can be eliminated.

A novel approach – that has only been possible since the move toward computer-based psychometrics – is the integration of warnings systems into personality measures. In this approach, candidates' response patterns are monitored in real-time. If an atypical pattern is detected – extreme responding that would suggest faking good or bad – a warning message is displayed to that candidate, reminding them that they should respond honestly. In real-world settings, this approach has proven very effective in curbing faking behaviour (Landers et al., 2011). However, it has been noted that accusations and negatively-worded warnings such as these – though more effective than positively-worded motivational messages – can increase candidates' test anxiety (Burns et al., 2015), suggesting that there may be ethical issues to consider around their use.

14.8   Online testing in specific contexts

An issue that invariably arises when discussing online versions of existing psychometric tools is the degree to which they are equivalent to their offline counterparts. It may seem that the medium in which a psychometric measure is presented would have little bearing on how it functions. However, a number of issues can affect how candidates respond to online measures versus their offline equivalents. Specific examples are discussed below, along with the issues associated with them.

14.9   Online testing for selection

A survey conducted by the Chartered Institute of Personnel and Development reports that approximately 24% of large organisations in the UK use some form of online psychometric testing to select job applicants (CIPD, 2015). Though this figure seems modest, it is worth considering in context: Approximately 47% of the organisations surveyed incorporate general ability testing into their selection processes, and 36% personality/aptitude questionnaires. It is reasonable to expect the prevalence of online testing to increase to approach these levels in the future, as more organisations realise the potential benefits of these tools over their traditional counterparts.

The potential for online measures to be remotely administered allows for them to be deployed more flexibly as part of selection processes. Traditional format measures require administration in proctored conditions, in which candidates are supervised by a test administrator. For this reason, it is only practical to use them to assess candidates as part of late-stage selection, when there are only a few candidates to be assessed. The removal of the need for measures to be administered face-to-face allows them to be used to assess candidates *en masse* as part of the early- to mid-stage of a selection process, typically as part of a second sift. This allows for much more robust prediction of future job performance and organisational fit than is typical at this stage of selection.

However, remote administration in selection contexts has brought with it a new problem: that of authentication. Even when candidates are provided with unique links or login details with which to identify themselves when completing assessments remotely, there is no guarantee that the person taking the test is who they say they are. Test publishers have begun to try to address this, often in novel ways. One solution is to monitor test takers remotely through a webcam, allowing the test proctor to establish authenticity and to detect cheating (Foster, 2009). However, this approach raises ethical issues around the candidate's right to privacy. One of the largest test publishers in the world, CEB SHL, take a somewhat different approach to this with their SHL Verify™ range of tests (SHL, 2013). In this approach, candidates first complete a remotely administered ability test. Those candidates who are successful at this stage later complete a shorter test, administered face-to-face, most frequently when attending the organisation's premises for interview. Comparable results across the two tests indicate that the results of the first, longer test are likely reliable. Vast differences in test performance would throw the reliability of the candidate's initial test result into question.

14.10 Online testing in training and development contexts

In training contexts, psychometric tests can be used as tools to analyse training needs and therefore develop more focused training regimens that suit both individual and organisational needs. In addition to using psychological assessments to define training needs, psychometric tests can be used to measure objectively the psychological constructs being developed pre-training, at the end of training, and as a post-programme follow-up to measure transfer of training or knowledge decay. For example, goals can be set for what is to be achieved in training, and training success can be measured against sets of psychological constructs, such as knowledge coherence and self-efficacy (for example Kozlowski et al., 2001).

Coaching contexts also benefit from use of psychometric tests. Measures of personality and well-being in particular are useful in coaching contexts. In addition to providing detailed information upon which to base personalised coaching interventions, psychometric tests can be used in similar ways to training interventions, measuring psychological constructs before and after the coaching intervention to compare performance on a given competency before, after, and at follow-up. Such evaluations can be useful in determining the utility and efficacy of coaching. This potential for psychometric tools to measure psychological characteristics both before and after an intervention is an advantage not just confined to employment settings. It also has applications in clinical and counselling, and forensic settings.

14.11   Clinical and therapeutic use of online assessment

Though covered extensively in other chapters, for the sake of completeness, brief mention must also be given to online psychological assessment in counselling and therapeutic contexts. Online counselling is an emerging area of practice (see Chapter 8), and can be less threatening to those who become anxious in traditional face-to-face settings. However, one must proceed with caution when using psychological assessment in online therapeutic contexts. Buchanan (2002) has raised concerns with online assessment of negative mood and emotion, stating that online versions of such assessments may not always be measuring the same constructs as their paper and pencil counterparts. Additionally, Buchanan warns that there may be ethical issues with using online assessment in counselling contexts rather than face-to-face versions, in that a therapist is less able to provide support to clients who experience distress as a result of the measure. Such issues must be given due consideration when using psychological assessments in online therapy, as the practitioner has a duty of care to ensure the well-being of the test taker/candidate, as well as the accuracy of the results.

14.12   Forensic and legal settings

Psychometric testing is becoming increasingly common in forensic and legal settings. Testing may be used for a range of applications to assess those accused of crime, offenders, and victims of crime. Psychological constructs measured include intellectual functioning, personality, emotional states, mental health and psychological well-being. In addition, assessments of literacy, numeracy, and even suggestibility are commonplace.

A forensic measure that has gained particular attention – both in the research literature and in media – is the Psychopathy Checklist-Revised (PCL-R; see Forth et al., 1996; Hare et al., 1990). It has been used to assess risk and measure treatment progress in a variety of forensic populations (Morrissey, 2003; Morrissey et al., 2007). Recently, a version of the PCL-R has been adapted for use in the workplace. B-Scan (Babiak & Hare, in preparation) is an online measure designed to detect corporate psychopathy, based on a combination of self-, supervisor, and peer ratings. Corporate psychopathy has received much research attention, as there is a growing body of evidence to suggest that individuals in leadership positions who display psychopathic traits can negatively impact on the job satisfaction and well-being of their employees (Mathieu et al., 2014), and are more likely to engage in corporate crime (Langbert, 2010). Corporate psychopathy has even been hypothesised as a causal factor in the 2008 Global Financial Crisis (Boddy, 2011). Tools such as B-Scan, therefore, allow practitioners to employ screening procedures to be better able to predict corporate crime, and to safeguard against management behaviours that could harm organisations and their staff.

Data from forensic psychological assessments are primarily used in making recommendations for mitigation and within sentencing. Therefore, it is essential to use theoretically sound, reliable, and valid measures. In addition, practitioners are responsible for ensuring that the testing environment is strictly controlled and monitored in addition to selecting appropriate tests for each individual (for example checking the reading level required and making reasonable adjustments that do not compromise the technical qualities of the test). Finally, the question of mistaken identity in the interpretation and use of testing results is less likely to arise face-to-face than in an online context. Therefore, while online psychological assessment for legal and forensic purposes is a possibility, the authors recommend caution when using them in such high-stakes contexts.

14.13   Careers guidance

The vocational psychologist has a range of tools available to aid in the process of giving careers guidance to a client. Personality measures may be used to identify a client's strengths, and to identify the types of job role to which these strengths would be an asset. Career interests inventories such as Holland's Self Directed Search (SDS; Holland et al., 1994) and the Strong Interests Inventory (SII; Hansen & Campbell, 1985) can be used in careers guidance to explore individuals' job preferences, motivations and interests, and match them to suitable working environments (O'Connell & Sedlacek, 1971; Savickas et al., 2002). Traditionally, these measures would uncover a client's job interests, and the practitioner would then be faced with the often-laborious task of searching for congruent environments by poring over Holland's Dictionary of Occupational Codes (Gottfredson & Holland, 1996).

Many online personality measures and interest inventories are now linked to data contained within O*NET (https://www.onetonline. org/), an online repository of occupational data maintained by the US Department of Labour/Employment and Training Administration. This allows practitioners to generate reports for clients instantly, substantially speeding up the process of career guidance using these methods. However, as in other contexts, psychometric measures cannot provide infallible careers advice, and practitioners should be discouraged from over-reliance on them. They should always be used in conjunction with the provision of thorough feedback, and as a starting point for a constructive relationship that benefits the client.

One particular innovation in how careers interest inventories are deployed is in the form of Ryopo™. Ryopo™ is based on Holland's Occupational Themes, a type-based model that categorises job interests and work environments as being either Realistic, Investigative, Artistic, Social, Enterprising, or Conventional, and is often, as a result, colloquially referred to as the RIASEC Model (Holland, 1997). Ryopo™ is a form of social network that functions as a matching service between job seekers and organisations (Ryopo, 2013). The system gathers information about a user's job interests, preferences, and work habits, then provides them with tailored search results, based on the congruence between these and work environments offered by employers who have paid to be listed on the site. Though relatively new, this form of career guidance is potentially paradigm-shifting: it may serve to automate a large proportion of the work involved in vocational psychology, and to provide it to a much wider audience than has previously been possible.

14.14 Conclusion

In summary, the proliferation of online psychometrics through many areas of psychological practice has brought with it many advantages over traditional paper-based formats. However, it has also presented a fresh set of challenges that need to be overcome. A fundamental issue that is often cited in the research literature is whether or not online and offline measures can truly be considered equivalent (Stone et al., 2015). Factors such as loss of control of the environment in which tests are administered, and the confounding effect of IT literacy and familiarity with computer equipment potentially introduce new sources of measurement error into the estimation of test takers' psychological characteristics when using remotely administered online measures.

Overall, though, there is cause for optimism: despite the long history of psychometric practice, its deployment as an online resource is still very much in its infancy. In years to come, psychometrics will continue to evolve as the technological landscape changes, allowing the problems of the past to be overcome, and allowing psychologists to measure the mind in ways that are impossible to predict.

14.15 Appendix

Table A14.1 Acronyms and initialisms

16PF	Cattell's Sixteen Personality Factors Questionnaire. This trait measure of personality, first developed in the 1940s, summarises personality as being composed of 16 traits. Traditionally seen as an alternative to Big Five measures, modern versions incorporate summary scales that map onto the Big Five personality traits.
BPS	The British Psychological Society. The UK's governing body for both academic and practitioner psychologists.
CAT	Computerised Adaptive Testing. A modern form of psychometric testing that adapts its content based on the responses of the test taker. Most frequently used for ability testing, CAT allows a test to be tailored to each individual candidate, reducing the time taken to estimate each one's true level of ability.
CEB SHL	Originally named Saville and Holdsworth Ltd. after its two founders, Peter Saville and Roger Holdsworth, SHL grew to be one of the largest psychometric testing publishers in the UK. In 2012, they were bought out by The Corporate Executive Board (CEB), a US advisory services company – to form CEB SHL.
CIPD	The Chartered Institute of Personnel and Development. The governing body of HR practice in the UK.

(continued)

250 Daniel Hinton and Debbie Stevens-Gill

Table A14.1   Continued

IM	Impression Management. First proposed as a facet of SDR, reflecting intentional, conscious attempts to create a more favourable impression of oneself than is actually the case. In modern usage, the two terms are, for the most part, used interchangeably.
MBTI	The Myers-Briggs Type Indicator. A hugely popular – yet somewhat controversial – type measure of personality that categorises respondents as belonging to one of 16 distinct personality types. Broadly based on the work of Carl Jung, these personality types are derived based on four dichotomous classifications: Extraversion/Introversion, Sensing/iNtuition, Thinking/Feeling, and Judgement/Perception. The resultant type is given a four-letter code to describe how a participant is classified on the dimensions, for example **ESTJ**; **INFP**.
NEO PI-R	Known colloquially as the NEO, this personality inventory is based on the Big Five Model. The NEO measures the traits of Neuroticism (vs. Emotional Stability), Extraversion, Openness, Conscientiousness, and Agreeableness. The first three of these give the NEO its name. The most common version of the NEO – the NEO PI-R – is the revised version of the inventory, first published in 1990.
O*NET	The Occupational Information Network. An extensive database of job information held and updated by the US Department of Labour/Employment and Training Administration.
PCL-R	The Psychopathy Checklist – revised. Designed for use in forensic samples, the PCL-R measures the personality traits and resultant behaviours most strongly associated with psychopathy. It measures facets such as lack of empathy, superficial charm, impulsivity, pathological lying, and promiscuous sexual behaviour, and tends to be administered by a training psychologist as part of a semi-structured interview.
RIASEC	A colloquial name for Holland's Occupational Themes, a model of job interests and work environments. The acronym RIASEC is derived from the six of occupational themes on which it is based (Realistic, Investigative, Artistic, Social, Entrepreneurial, and Conventional).
SDR	Socially Desirable Responding behaviour. A form of response bias, SDR is a set of behaviours that suggest a respondent is presenting a more favourable impression of him or herself than is actually the case, for example appearing more conscientious and hardworking than they are in reality.
SDS	The Self-Directed Search. A type-based career interest inventory based on Holland's Occupational Themes. The SDS is used to aid careers guidance by matching respondents to work environments, based on the degree to which their job interests reflect aspects of these environments.

(continued)

Table A14.1 Continued

SII	The Strong Interest Inventory. First developed by Edward Strong, the modern version of this career interest inventory is based on Holland's Occupational Themes. It is seen as an alternative measure to the SDS, and is used in much the same way.
SJT	The Situational Judgement Test. An approach to measuring the knowledge and/or behaviour of a test taker, relevant to a specific work context. Participants are presented with a hypothetical scenario – either as text or as video – and asked to select a response from a list of possible reactions to that scenario. Depending on how the scenario is presented, test takers either indicate how they *should* respond, or how they *would* respond.
WAIS	The Wechsler Adult Intelligence Scale. A battery of ability measures, used to estimate the general mental ability (expressed in terms of IQ) of adult test takers. It is currently on its fourth iteration, the WAIS-IV.
WISC	The Wechsler Intelligence Scale for Children. First released around the same time as the WAIS, the WISC measures general mental ability in children, adjusting its resultant measure of IQ based on a test taker's age.

14.16 References

Ackerman, P. L., Kanfer, R., & Goff, M. (1995). Cognitive and non-cognitive determinants and consequences of complex skill acquisition. *Journal of Experimental Psychology, 1*, 270–304.

Armon, G., Shirom, A., & Melamed, S. (2012). The big five personality factors as predictors of changes across time in burnout and its facets. *Journal of Personality, 80*(2), 403–427.

Arthur, W., Glaze, R. M., Villado, A. J., & Taylor, J. E. (2010). The magnitude and extent of cheating and response distortion effects on unproctored internet-based tests of cognitive ability and personality. *International Journal of Selection and Assessment, 18*(1), 1–16.

Babiak, P., & Hare, R. D. (in preparation). *The B-Scan 360 Manual.*

Back, M. D., Stopfer, J. M., Vazire, S., Gaddis, S., Schmukle, S. C., Egloff, B., & Gosling, S. D. (2010). Facebook profiles reflect actual personality, not self-idealization. *Psychological Science, 21*(3), 372–374.

Bergman, M. E., Drasgow, F., Donovan, M. A., Henning, J. B., & Juraska, S. E. (2006). Scoring situational judgment tests: Once you get the data, your troubles begin. *International Journal of Selection and Assessment, 14*(3), 223–235.

Bjornsdottir, G., Almarsdottir, A. B., Hansdottir, I., Thorsdottir, F., Heimisdottir, M., Stefansson, H., Thorgeirsson, T. E., & Brennan, P. F. (2014). From paper to web: Mode equivalence of the ARHQ and NEO-FFI. *Computers in Human Behavior, 41*, 384–392.

Boddy, C. R. (2011). The corporate psychopaths theory of the global financial crisis. *Journal of Business Ethics, 102*(2), 255–259.

Bretz, R. D., & Judge, T. A. (1994). Person–organization fit and the theory of work adjustment: Implications for satisfaction, tenure, and career success. *Journal of Vocational Behavior, 44*(1), 32–54.

Brown, D. V. (2006). *Tricks of the Mind*. London: Channel 4 Books.

Buchanan, T. (1999). Online personality assessment: Equivalence of traditional and WWW personality measures. *German Online Research, 99*. Retrieved on 10/7/2015, from http://gor.de/archive/images/gor99/tband99/pdfs/a_h/buchanan.pdf

Buchanan, T. (2002). Online assessment: Desirable or dangerous? *Professional Psychology: Research and Practice, 33*(2), 148–154.

Burns, G. N., Fillipowski, J. N., Morris, M. B., & Shoda, E. A. (2015). Impact of electronic warnings on online personality scores and test-taker reactions in an applicant simulation. *Computers in Human Behavior, 48*, 163–172.

Carroll, J. B. (1993). *Human Cognitive Abilities: A Survey of Factor-analytic Studies*. Cambridge: Cambridge University Press.

Chan, D., & Schmitt, N. (1997). Video-based versus paper-and-pencil method of assessment in situational judgment tests: Subgroup differences in test performance and face validity perceptions. *Journal of Applied Psychology, 82*(1), 143–159.

Chartered Institute of Personnel and Development. (2015). *Survey Report: Resourcing and Talent Planning 2015*. Retrieved on 17/6/2015 from https://www.cipd.co.uk/binaries/resourcing-talent-planning_2015.pdf

Christian, M. S., Edwards, B. D., & Bradley, J. C. (2010). Situational judgment tests: Constructs assessed and a meta-analysis of their criterion-related validities. *Personnel Psychology, 63*(1), 83–117.

Chuah, S. C., Drasgow, F., & Roberts, B. W. (2006). Personality assessment: Does the medium matter? No. *Journal of Research in Personality, 40*(4), 359–376.

Coaley, K. (2014). *An Introduction to Psychological Assessment and Psychometrics* (2nd edition). London: Sage.

Conn, S. R., & Rieke, M. L. (1994). *16PF Fifth Edition Technical Manual*. Savoy, IL: Institute for Personality & Ability Testing, Inc.

Costa, P. T., & McCrae, R. R. (1992). *NEO PI-R. Professional Manual*. Odessa, FL: Psychological Assessment Resources.

DeBerard, M. S., Spielmans, G., & Julka, D. (2004). Predictors of academic achievement and retention among college freshmen: A longitudinal study. *College Student Journal, 38*(1), 66–80.

Forer, B. R. (1949). The fallacy of personal validation: a classroom demonstration of gullibility. *Journal of Abnormal and Social Psychology, 44*(1), 118–123.

Forth, A. E., Brown, S. L., Hart, S. D., & Hare, R. D. (1996). The assessment of psychopathy in male and female noncriminals: Reliability and validity. *Personality and Individual Differences, 20*(5), 531–543.

Foster, D. (2009). Secure, online, high-stakes testing: Science fiction orbBusiness reality? *Industrial and Organizational Psychology, 2*(1), 31–34.

Furnham, A., Eracleous, A., & Chamorro-Premuzic, T. (2009). Personality, motivation and job satisfaction: Hertzberg meets the Big Five. *Journal of Managerial Psychology, 24*(8), 765–779.

Goldberg, L. R. (1990). An alternative 'description of personality': The Big-Five factor structure. *Journal of Personality and Social Psychology, 59*, 1216–1229.

Gosling, S. D., Augustine, A. A., Vazire, S., Holtzman, N., & Gaddis, S. (2011). Manifestations of personality in online social networks: Self-reported

Facebook-related behaviors and observable profile information. *Cyberpsychology, Behavior, and Social Networking, 14*(9), 483–488.

Gosling, S. D., & Mason, W. (2015). Internet research in psychology. *Annual Review of Psychology, 66,* 877–902.

Gottfredson, G. D., & Holland, J. L. (1996). *Dictionary of Holland Occupational Codes.* Odessa, FL: Psychological Assessment Resources.

Harackiewicz, J. M., Barron, K. E., Tauer, J. M., & Elliot, A. J. (2002). Predicting success in college: A longitudinal study of achievement goals and ability measures as predictors of interest and performance from freshman year through graduation. *Journal of Educational Psychology, 94*(3), 562–575.

Hare, R. D., Harpur, T. J., Hakstian, A. R., Forth, A. E., Hart, S. D., & Newman, J. P. (1990). The revised Psychopathy Checklist: Reliability and factor structure. *Psychological Assessment: A Journal of Consulting and Clinical Psychology, 2*(3), 338–341.

Hodson, G., & Busseri, M. A. (2012). Bright minds and dark attitudes lower cognitive ability predicts greater prejudice through right-wing ideology and low intergroup contact. *Psychological Science, 23*(2), 187–195.

Hansen, J. L. C., & Campbell, D. P. (1985). *Manual for the Strong Interest Inventory.* Palo Alto, CA: Consulting Psychologists Press.

Holland, J. L. (1997). *Making Vocational Choices: A Theory of Vocational Personalities and Work Environments.* Odessa, FL: Psychological Assessment Resources.

Holland, J. L., Powell, A. B., & Fritzsche, B. A. (1994). *The Self-Directed Search (SDS).* Odessa, FL: Psychological Assessment Resources.

Kantrowitz, T. M., Dawson, C. R., & Fetzer, M. S. (2011). Computer adaptive testing (CAT): A faster, smarter, and more secure approach to pre-employment testing. *Journal of Business and Psychology, 26*(2), 227–232.

Kozlowski, S. W., Gully, S. M., Brown, K. G., Salas, E., Smith, E. M., & Nason, E. R. (2001). Effects of training goals and goal orientation traits on multidimensional training outcomes and performance adaptability. *Organizational Behavior and Human Decision Processes, 85*(1), 1–31.

Landers, R. N., Sackett, P. R., & Tuzinski, K. A. (2011). Retesting after initial failure, coaching rumors, and warnings against faking in online personality measures for selection. *Journal of Applied Psychology, 96*(1), 202–210.

Langbert, M. B. (2010). Managing psychopathic employees. *Cornell HR Review.* Retrieved on 9/7/2015 from http://digitalcommons.ilr.cornell.edu/chrr/1

Lievens, F., Buyse, T., & Sackett, P. R. (2005). The operational validity of a video-based situational judgment test for medical college admissions: Illustrating the importance of matching predictor and criterion construct domains. *Journal of Applied Psychology, 90,* 442–452.

Lievens, F., Decaesteker, C., Coetsier, P., & Geirnaert, J. (2001). Organizational attractiveness for prospective applicants: A person–organisation fit perspective. *Applied Psychology, 50*(1), 30–51.

Mathieu, C., Neumann, C. S., Hare, R. D., & Babiak, P. (2014). A dark side of leadership: Corporate psychopathy and its influence on employee well-being and job satisfaction. *Personality and Individual Differences, 59,* 83–88.

Mezquita, L., Stewart, S. H., & Ruipérez, M. Á. (2010). Big-five personality domains predict internal drinking motives in young adults. *Personality and Individual Differences, 49*(3), 240–245.

Morey, L. C., & Quigley, B. D. (2002). The use of the Personality Assessment Inventory in assessing offenders. *International Journal of Offender Therapy and Comparative Criminology, 46,* 333–349.

Morrissey, C. (2003). The use of the PCL-R in forensic populations with learning disability. *The British Journal of Forensic Practice, 5*(1), 20–24.

Morrissey, C., Mooney, P., Hogue, T. E., Lindsay, W. R., & Taylor, J. L. (2007). Predictive validity of the PCL-R for offenders with intellectual disability in a high security hospital: Treatment progress. *Journal of Intellectual and Developmental Disability, 32*(2), 125–133.

Myers, I. B., McCaulley, M. H. K., Quenk, N. L., & Hammer, A. L. (1998). *MBTI manual: A guide to the development and use of the Myers-Briggs Type Indicator* (Vol. 3). Palo Alto, CA: Consulting Psychologists Press.

O'Connell, T. J., & Sedlacek, W. E. (1971). *The Reliability of Holland's Self Directed Search for Educational and Vocational Planning.* Retrieved on 10/7/2015, from http://files.eric.ed.gov/fulltext/ED065524.pdf

Ortner, T. M. (2008). Effects of changed item order: A cautionary note to practitioners on jumping to computerized adaptive testing for personality assessment. *International Journal of Selection and Assessment, 16*(3), 249–257.

Pearson Education Ltd. (2015a). *Raven's APM and SPM Short-Forms.* Retrieved on 25/6/2015 from http://www.talentlens.co.uk/select/ravens-apm-and-spm-short-forms

Pearson Education Ltd. (2015b). *Wechsler Adult Intelligence Scale®- Fourth UK Edition.* Retrieved on 25/6/2015 from http://www.helloq.co.uk/overview/the-qinteractive-library/wais-iv.html

Pearson Education Ltd. (2015c). *Wechsler Intelligence Scale for Children®- Fourth Edition.* Retrieved on 25/6/2015 from http://www.helloq.co.uk/overview/the-qinteractive-library/wisc-iv.html

Peeters, H., & Lievens, F. (2005). Situational judgment tests and their predictiveness of college students' success: The influence of faking. *Educational and Psychological Measurement, 65*(1), 70–89.

Penney, L. M., David, E., & Witt, L. A. (2011). A review of personality and performance: Identifying boundaries, contingencies, and future research directions. *Human Resource Management Review, 21*(4), 297–310.

Ree, M. J., & Carretta, T. R. (1998). General cognitive ability and occupational performance. In C. L. Cooper & I. T. Robertson (Eds.), *International Review of Industrial and Organizational Psychology* (Vol. 13, pp. 159–184). Chichester, England: Wiley.

Ree, M. J., & Earles, J. A. (1991). Predicting training success: Not much more than g. *Personnel Psychology, 44,* 321–332.

Robertson, I., & Smith, M. (2001). Personnel selection. *Journal of Occupational and Organizational Psychology, 74,* 441–472.

Rosenberg, J., & Egbert, N. (2011). Online impression management: Personality traits and concerns for secondary goals as predictors of self-presentation tactics on Facebook. *Journal of Computer-Mediated Communication, 17*(1), 1–18.

Ryopo Ltd. (2013). *Ryopo: About Us.* Retrieved on 25/6/2015 from http://ryopo.com/about

Savickas, M. L., Taber, B. J., & Spokane, A. R. (2002). Convergent and discriminant validity of five interest inventories. *Journal of Vocational Behavior, 61*(1), 139–184.

Schmidt, F. L., & Hunter, J. E. (1998). The validity and utility of selection methods in personnel psychology: Practical and theoretical implications of 85 years of research findings. *Psychological Bulletin, 124*(2), 262–274.

SHL. (2013). *Aptitude: Identify the Best Talent Faster and at Less Cost.* Retrieved on 25/6/2015 from http://ceb.shl.com/assets/Aptitude-brochure-UKeng.pdf

Sjöberg, L. (2015). Correction for faking in self-report personality tests. *Scandinavian Journal of Psychology, 56*(5), 582–591.

Spector, P. E., & O'Connell, B. J. (1994). The contribution of personality traits, negative affectivity, locus of control and Type A to the subsequent reports of job stressors and job strains. *Journal of Occupational and Organizational Psychology, 67*(1), 1–12.

Stone, D. L., Deadrick, D. L., Lukaszewski, K. M., & Johnson, R. (2015). The influence of technology on the future of human resource management. *Human Resource Management Review, 25*(2), 216–231.

Triantafillou, E., Georgiadou, E., & Economides, A. A. (2008). The design and evaluation of a computerized adaptive test on mobile devices. *Computers & Education, 50*(4), 1319–1330.

van der Linden, W. J. (2008). Some new developments in adaptive testing technology. *Zeitschrift für Psychologie/Journal of Psychology, 216*(1), 3–11.

van der Linden, W. J. (2009). Predictive control of speededness in adaptive testing. *Applied Psychological Measurement, 33*(1), 25–41.

Weiss, B., & Feldman, R. S. (2006). Looking good and lying to do it: Deception as an impression management strategy in job interviews. *Journal of Applied Social Psychology, 36*(4), 1070–1086.

Wonderlic, Inc. (2014). *Wonderlic Personnel Test (WPT-R).* Retrieved on 25/6/2015 from http://www.wonderlic.com/assessments/ability/cognitive-ability-tests/contemporary-cognitive-ability-test

Woods, S. A., & Hampson, S. E. (2010). Predicting adult occupational environments from gender and childhood personality traits. *Journal of Applied Psychology, 95*(6), 1045–1057.

Yang, B., Kim, Y., & McFarland, R. G. (2011). Individual differences and sales performance: A distal-proximal mediation model of self-efficacy, conscientiousness, and extraversion. *Journal of Personal Selling & Sales Management, 31*(4), 371–381.

Index